W9-BKU-100

jane grosslight

illustrations by jeffrey w. verheyen

effective use of daylight and electric lighting in residential and commercial spaces

LIGHT
LIGHT
LIGHT

DURWOOD PUBLISHERS
BOX 37474 TALLAHASSEE, FL 32315

First edition published by Prentice Hall,
Sprectrum Division, 1984
Library of Congress Cataloging in Publication Data
Grosslight, Jane
    Light, Light, Light, effective use of daylight and
elctric lighting in residential and commercial spaces.
    Bibliography: p.
    Includes index, glossary.
    1. Dwellings—Lighting. 2. Commercial buildings—
Lighting.
    3. Daylight 4. Electric lighting I. Title.

Second edition, revised and published
by Durwood Publishers, 1990.

Third edition, revised and published
by Durwood Publishers, 1998.

10   9   8   7   6   5   4

This book is available at a special discount when
ordered in bulk quanatities. Contact Durwood
Publishers.

Illustrations by Jeffrey W. Verheyen
Original cover concept by Hal Siegel
Original book design and page layout by Maria Carella

Originally published by
Prentice-Hall, Inc
Englewood Cliffs NJ

Republished by
Durwood Publishers
Box 37474
Tallahassee FL
32315-7474

ISBN 0-927412-06-3

Notice of Liability
The information contained in this book is distributed
without warranty. Neither the author nor the publisher
shall have any liability to any person or any entity for
any loss or any damaged alleged to be caused or caused,
either directly or indirectly by the information contained in this book. The author and publisher caution
that all lighting components need to be installed by a
licensed electrical contractor and in compliance with
all applicable codes.

contents

foreword

Everyone appreciates good lighting design. We can all identify lighting environments that are a delight to the senses, and we also know when lighting is less than satisfactory. As clients and users become literate in lighting design, the entire design profession benefits. The dialogue between designer and client improves as our expectation for better environments rises.

Architects and interior designers have always understood the importance of light and shadow in revealing their work. They strive to achieve lighting effects that are in harmony with the building. Lighting must be visually comfortable and safe and meet both energy and cost constraints by using the best available technology. But technology alone cannot give superior results. Lighting designers use intuition and artistic imagination to design spaces that inspire and stimulate the moods of those who work or live there. Improved lighting environments have both physical and emotional rewards; human spirits are lifted by a sparkle of daylight in the right place.

This book helps everyone understand both the tangible and intangible benefits of good lighting. It should appeal to a wide audience; students of lighting design will find solutions to many typical problems encountered in residential and commercial practice. For those who simply wish to better understand their own lighting environments, lighting design is described in a profusion of illustrations and examples.

Professionals, clients, students and others who are interested in healthy, lively, well-lighted spaces will find this book fascinating and valuable. After reading it, I think you will agree that there are no mysteries to good lighting design.

David Lord
Professor of Architecture
California Polytechnic State University

preface to 1998 version

Lighting has dramatically and continually changed since this book was first published in 1984. Primarily, lighting has been affected by federal and state legislation mandating energy efficiency. Thus, many light sources have been outlawed. Concomitantly, manufacturers have developed new sources and fixtures, and modified existing ones to comply with the mandates. This edition, therefore, required taking out unavailable sources that we have all used in the past and adding the many new sources and their applications. The choices now are more numerous than before, but the differences between choices are more subtle and demanding. This complexity does not have to be perplexing. This book intends to inform.

The readers of this book have changed over the years from end-users to professional architects and interior designers and students of design. Hence, the focus has changed from possibly-doing-it-yourself to doing-it-for-others. This third edition targets the professional audience with information in a concise manner.

Originally, I envisioned the book in college classrooms for teachers wanting application information as well as technology, and in libraries across the country being a reference for anyone. Those goals were met. But now more and more professionals keep it handy as a reference for designing lighting. Some architects give it to their clients before tackling the lighting plan.

The use of residential space has changed. Home-based businesses now demand professional-quality office lighting suitable for both paper-based and screen-based visual tasks. Hence, information contained in Chapter 19 (The Best Possible Desk Lighting) should be utilized in home spaces.

More than 100,000 residential units under construction this year will have lighting installed. Occupants of already-built residences can relight to suit their life-style, to brighten their spaces, and to increase their home's value. Thus, the potential number of residential lighting clients is almost unlimited.

Small commercial spaces (300 to 3,000 square feet) line our highways. Some are boutiques, beauty salons, or bakery shops. Others offer insurance, income tax assistance, or investment advice. All offer a product or a service. All can improve their business through lighting without a big investment and with a long-range payback if energyefficient. Typically, every five years retail stores totally redesign their lighting— sometimes corporate legislated; sometimes market position demanded. Nonetheless, the potential number of commercial lighting clients is vast.

Off-the-shelf luminaires may be found in lighting catalogs and in lighting showrooms depending upon the availability in your area. About 200 manufacturers make luminaires and around 2,000 lighting showrooms display them. But the distribution system is imperfect. Accessibility to all the choices is difficult; lighting consultants can provide the greatest choices.

This third edition, like the first two, provides line drawings showing what light does (in a stylized manner) to help with decisions. The intention is to be instructional for easy light-source and lighted-surface identification. Photographs do not convey such information as concisely.

This book is a guide for lighting design. Many choices for lighting residential and small commercial spaces are included. Practical know-how, up-to-date information, and how-to directions are provided. Terms are defined in the illustrated glossary. The index is extensive. Detailed directions for custom-made applications are included. Rules of Thumb are provided to simplify initial designs. Sample Electric Costs compare long-range costs of different lighting decisions. Differences that impinge upon lighting decisions are included and alternatives for changes are supplied. The book's scope ranges from necessities to luxuries and for lighting inside and out. Use it for selected readings now and as a reference later.

light
for spaces

Design the best possible lighting by defining the lighting problem well and understanding the potential solutions. Usually, one solution is not the only appropriate solution. Thus, an extensive amount of information must be available so that many solutions may be examined.

In residential spaces, light is needed at night and during the day if gloomy outside or the window is overly bright. Lighting should make visual tasks visible and should make the space pleasing, colorful, and flattering. Always redesign residential lighting if redecorating or relocating. Always design the best possible lighting for commercial spaces to improve productivity, stimulate sales, or create a conducive environment. Light has powerful effects; use it to its fullest advantage!

Light is important. Without it, form and colors cannot be seen and functions cease. Remember when the lights went out? If you did not have a candle, you probably went out also, or went to bed. Most people take light for granted and endure whatever amount they receive—good, bad, or indifferent—from whatever luminaires deliver it. Unfortunately, lighting is rarely designed to be the best possible. Luminaires (lighting fixtures) attached to a structure are usually determined by the builder or the electrical subcontractor and chosen because of least initial cost rather than good output and low, long-term operating cost. In residences, occupants add portable luminaires (table lamps) chosen for decorative qualities rather than light-delivering capabilities. In the same way, occupants of commercial offices add personal desktop luminaires, retrofit ceiling downlights without regard for quality of light, or they struggle under the poor light. However, lighting can be designed to be the best possible.

Reprinted by permission of King Features Syndicate, Inc.

The best possible lighting is like music; it enhances tasks and gives pleasure. Light can enclose, direct, and flatter. Poor lighting can be too much and, like noise, detract and irritate. Or, poor lighting can be too little, making visual tasks difficult. Or it can create a gloomy space deprived of color, highlight, and shadow. Not good.

Spaces can be well lighted without increasing the utility bill if daylight is used effectively and if electric lighting is well designed and used judiciously. Daylight is often not taken into consideration as a light source. It should be. It is free light, as long as it does not also bring in unwanted heat in hot climates or bring in debilitating glare. Electric light can be energy-efficient and used judiciously.

The lighting recommendations in this book are tailored to many differences—personal, structural, and visual tasks. Personal differences include economics, age, life-style, and ownership. Economics dictates how much money is available for lighting. The best possible lighting need not be expensive, just well planned. People of different ages have different requirements for light. And the same-age people have different lighting preferences—brighter or softer. Likewise, life-style affects lighting choices—cooks need more light in the kitchen than noncooks. Ownership affects lighting considerations. On the one hand, owners may recess their lighting. On the other hand, renters must not. Residential renters may want to remove their luminaires when they move. Commercial renters often must leave installed luminaires behind. The constraints of short-term and mobile-home owners make their lighting solutions similar to those of renters. Likewise, owners planning to build a new structure have more choices than owners remodeling. All things need to be considered.

Structural differences include structure type and characteristics of the space. Such differences affect what kind of luminaires can be used, how the light is received and reflected, and where the luminaires can be placed.

Finally, visual task differences include task characteristics that affect the amount of light needed. For example, critical tasks require more light than casual tasks; small details require more light than large details; and short-viewing time requires more light than long-viewing time. Lighting design must be tailored to these differences.

Contrary to typical thinking, professional lighting design does not start with decisions about the light fixture. The first decision is what is to be illuminated, the second is which *light source* can do it, and the third is what fixture can be used.

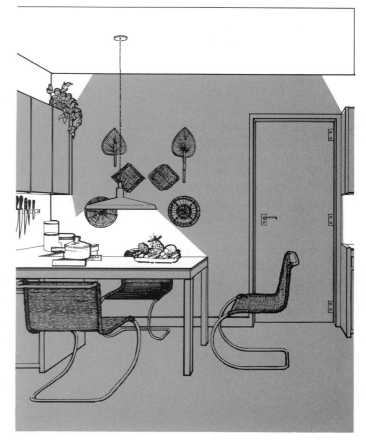

Light your space without lighting up your electric bill.

Use daylight effectively.

who should design lighting?

Lighting Consultants

Lighting design combines science and art. It is not simple. Most lighting has not been skillfully designed. Imitating what is seen most often repeats poor lighting. Lighting consultants can provide skillfully designed lighting. They keep up with the technological changes and latest products. Using them as consultants is a sensible solution to technology overload. Lighting consultants can specify lighting systems and controls. They can save money in energy-use and light-source-replacement costs over the life of the installation. Lighting consultants can be:

- independent lighting consultants or lighting designers
- architects and interior designers well-versed in lighting
- experienced lighting showroom staff or manufacturers' reps
- lighting-management contractors and energy managers

On the one hand, showroom staff and reps may be knowledgeable about the products that they sell and often do not charge for design services, but they ARE selling their products. On the other hand, independent lighting consultants have no financial interests in any particular product, specify the best lighting for the job, and must charge for design services.

If you wish to use a lighting consultant's highly skilled services, agree on what is expected to be done and who makes which decisions. Have furniture layout and colors chosen. (The more information given, the better the final result.) If you do not know what you want, tell your consultant. Consultants have many excellent design ideas; you can choose among them. The illustrations in this book will help you visualize how a design will look and may inspire more ideas.

After choosing a consultant, check to make sure that the cost and method of payment for the services are specified and that financial limits are clearly defined for the lighting job. The design process is a sequence of tasks for which the client, the consultant, or both are responsible.

Some state regulations limit lighting design for certain building types to state-registered professionals who do not specialize in lighting. They specialize in other aspects of design. Although well intended, the regulations do not guarantee

TASKS AND RESPONSIBILITIES FOR LIGHTING DESIGN

	CLIENT	CONSULTANT	JOINT
Discussion and job definition			√
Contract			√
Design analysis		√	
Design proposal			√
Conference on lighting design proposal			√
Selection of design	√		
Preliminary estimates		√	
Conference on estimates			√
Revision if necessary		√	
Selection of electrician	√	or	√
Final lighting plan and fixture schedule		√	
Acceptance of final plan	√		
Awarding electrical contract		√	
Ordering lighting fixtures	√	or	√
Scheduling installation	√	or	√
Inspecting and approving installation	√	or	√
Installation revision, if necessary			√
Lamping and focusing fixtures			√
Evaluation of installation			√
Final evaluation	√		

Adapted from *How to Market Professional Design Services* by Gerre Jones,
© 1973, McGraw-Hill Book Co., New York.

the quality of lighting design able to be provided, because the professionals enumerated often do not have time to keep up with lighting technology and the other technologies. Truly, lighting is a highly specialized field. Hence, certification of lighting consultants has been established. Those certified are able to use LC (Lighting Certification) appellation.

Many, Many Choices

Contrary to commonly held beliefs, lighting is rich with design choices. Different choices yield different effects and uses.

See for Yourself: How to Light This Room

1. With no constraints of money, installation, or client's style, how many ways can you identify to light the room illustrated? (If you drew the sun or the moon outside the window, you get an "A" for this lesson. Indeed, both yield light.)

2. At the least, electric lights can be in four places: bed, dressing table, chaise lounge, and outside patio. Minimally, twenty-five applications are possible:

    At the bed
    wall wash
    wall-mounted reading luminaire
    up/down bracket above bookshelf
    cornice inside canopy
    cove uplight on top of canopy
    low-voltage tubes in niche
    aqua-colored cold cathode in niche
    low-voltage strips around the base of bed

electroluminescent strips around the bed

At the dressing table
recessed, louvered, fluorescent downlight in soffit
2 incandescent downlights aimed at an angle in soffit
2 wall lanterns on either side of mirror
surface-mounted, low-voltage, G-lamps beside the mirror

At the chaise lounge
recessed downlight
pendant downlight
pendant up/downlight
downlight in a chandelier
floor lamp
uplight valance over window wall

On the patio
wall lamp to match inside wall lamps
uplight at potted evergreen
floodlight at top of the interior wall
3 recessed downlights on roof overhang
low-voltage string lights in the potted evergreen
grazing light on brick wall

What an array! Most often, bedrooms have a center ceiling light pumping 150 incandescent watts (causing shadows and glare) and a dresser or a bedside portable luminaire of 100 incandescent watts. However, the total (250 watts) could be spread around the room in different light sources and be more effective and more energy efficient. Spread the watts around.

The first illustration has enough light at the mirror for inspection but not for makeup. The third has enough light at the bed for short-term but not long-term reading. Tailor the lighting design to the needs of the client and use the many applications available. As long as it is not glaring, you cannot have too many light sources. Make lighting the best possible.

At the bed: 46 watts of T-2 fluorescent in canopy.
At the dressing table: two 15-watt sources in each wall luminaire.
At the chaise: six 5-watt low-voltage in chandelier and 20-watt PAR-20 downlight.
On the patio: two 15-watt sources in wall luminaire.

The total is 186 watts.

At the bed: 9-watt compact fluorescent reading light.
At the dressing table: two 32-watt (non-hazardous disposal) fluorescent recessed.
At the chaise: a 25-watt compact fluorescent floor luminaire.
On the patio: one 20-watt R compact fluorescent in uplight.

The total is 118 watts.

At the bed: two 20-watt, MR-16 wall-washers.
At the dressing table: ten low-voltage, 2.5-watt G sources beside the mirror.
At the chaise: one 15-watt, G compact fluorescent.
On the patio: one 15-watt R compact fluorescent in floodlight.

The total is 95 watts.

what you need to know

2

An extensive amount of information is required to design the best possible light in a suitable amount from the best direction in the best color to enhance visibility. Much needs to be known!

verbally describe lighting

Your clients' preferences need to be known. Have your clients choose adjectives describing their style. Do they want the lighting to be sophisticated or simple? There is a difference. Do they want it to be opulent or rustic? The adjectives can become an informal contract between you and the client to measure the design (discussions and end product). Your interpretation and theirs can be adjusted and matched so that everyone is happy with the results. Adjectives that can describe lighting are:

sophisticated .... simple
stimulating .... restful
distinctive .... conventional

robust .... delicate
bold .... subtle
historic .... contemporary
opulent .... rustic
custom-made .... off-the-shelf
formal .... informal
dramatic .... casual
luxurious .... economical
classic .... trendy
frivolous .... frugal

You need to know what your clients prefer. (I had a client whom I assumed would want delicate and sophisticated lighting; she indicated that she liked rustic and bold. Oh, such good information to have early on!)

design visually

How the light will look needs to be known. Light is not a solid object. Therefore, do not design lighting on a reflected ceiling plan. It is not effective. Such a plan communicates with the installer, but, it does not communicate to you during the design process. It certainly does not communicate with your clients who, for the most part, cannot visualize three dimensionally. Design lighting on a one- or two-point perspective view.

Structures and furniture are solid objects; they can be designed on a plan view—although the result is better if they are also designed in perspective.

If you design lighting on a reflected ceiling plan first, you will continually create what you know over and over again—cookie-cutter lighting. You will lose any chance of discovery. Opportunities for lighting that are not obvious in a plan view can be revealed in a perspective drawing.

How to design visually? Use quick sketches; cut and paste reprographics; tracings of enlarged Polaroid prints from already-built spaces; marker design renderings; computer-aided perspectives (CAD); Virtual Reality; or the "talking paper"—Rub & Show—from this publisher that has an erasable, renewable surface to show lighting schemes (looks like the illustrations in this book). Whatever method, play with what the light will look like, again and again. Don't forget lighting is forever visual.

where is light seen?

Where is light seen in a space? Light is seen only in two places: first, on surfaces, and second, at a visible light source, if any. Nowhere else! Look around. A lighted surface—tabletop, wall, floor, or reading material—reflects light and can be seen. Light is invisible in the air; it cannot be seen.

Further, if a bare light source is lit and can be viewed (a chandelier without shades), light is seen. Light on a lens of a fluorescent ceiling luminaire is a surface reflecting light, not a source. Yes, it is translucent and at certain angles a source can be seen through it. And yes, it is bright and the bright lens has a profound effect on the visual impact of the lighting, but nonetheless is a surface reflecting light, not a source. Light can only be seen directly at a source or on a surface. Think of the sun (light at a source) and the moon (a lighted surface).

When designing, answer this question: "Do you want your client to see bare light sources and what surfaces do you want to light?"

A bare (low-wattage) light source can add sparkle to a space—but that's the icing on the cake—a nicety. Lighted task surfaces are the meat and potatoes—the necessities. Whether a desktop, a face in a mirror, a newspaper being read, or a piece of wood in a vise, light is necessary. But lighted surfaces can also be a nicety—accented art, a grazed textured wall, or an edge-lit etched-glass divider.

Know what sources you want seen and what surfaces you want to light. Then, determine how much light and from what direction.

how much light?

Our eyes are exceedingly adjustable; they adjust to both moonlight and bright sunlight. Rarely do they pain us. But, abrupt eye adjustment can cause short-term pain—for example, going quickly from a dark area to a bright area, such as coming into the sunlight after a matinee movie. Fully dark-adapted eyes do not adjust quickly to brightness. Usually the effects of bad lighting are subtle and occur almost unnoticeably (a sense of fatigue) or later (as a headache or tension in the neck and shoulders). Sometimes the effects can be eyestrain from having to adjust repeatedly to brightness, darkness, brightness, darkness.

The brain has difficulty judging the actual amount of light seen and compares the previous amount. Thus, if it was bright and now is not, the light is judged to be dim. If it was dim and still is, the eyes adapt and the light is judged sufficient unless it fails to satisfy our needs.

Likewise, psychological overtones affect our judgement of light. For example, small areas of bright light create an impression of overall brightness, particularly with uplight distributions, even if the total amount is not greater. Hence, office-pendant-uplights have pinhole apertures along the sides of pendants to create a greater perceived brightness. Further, walls washed with soft light give a perception of greater room brightness.

How much light is needed is not just a matter of amount. It is a matter of visibility. Visibility is dependent upon amount and direction of light, and size, contrast, and background of the task. The amount has to be appropriate to illuminate the task. The better the contrast between the details of the visual task and its background, the less light is needed. A suitable direction to reveal contrast and details is required. From the wrong direction, light can obscure details even in a correct amount. Less light from the right direction could produce greater visibility.

In addition, how much light is needed depends upon how good our eyes are and what speed and accuracy are required for the task. Age affects eyes. People over 50 require more light to see well. Speed is determined by how quickly the task must be completed. And accuracy may be critical. Clearly, reading a Gothic novel for pleasure is not critical. But, proofreading a Gothic novel is.

The Illuminating Engineering Society of North America (IESNA) has developed a range of illumination amounts needed for activities measured in footcandles (amount of light falling on a surface). The range is based on the viewer's age, the reflectance value of the task's background, and the speed and accuracy demands.

The reflectance value is determined by surface colors. Dark colors reflect little light; pale colors reflect more light, but never 100 percent—consult technical paint sample charts for the light reflectance percentage of colors.

Generally in residences, casual activities—everyday living and moving about the space—require up to 10 footcandles of light. Easy tasks—reading and preparing food—require up to 50 footcandles. Extended tasks—hobby work and

Adapted from Osram Sylvania.

prolonged reading—require up to 150 footcandles. Difficult tasks—sewing and color matching—require up to 200 footcandles. If, for instance, the eyes are older, the reflectance value is low, and speed is required, the highest number of footcandles should be used. If the eyes are young, the reflectance value is high, and accuracy is unimportant, a lower number of footcandles is acceptable.

Generally, in commercial spaces, circulation can require up to 15 footcandles. Depending upon the type of business, merchandise appraisal requires up to 75 footcandles. Feature displays can require up to 500 footcandles to be distinguished from their surroundings. Difficult visual activities, such as drafting, require 200 or more footcandles. (Consult IESNA *Lighting Handbook, Reference and Application*, 1993.)

Footcandles are lost between the source and the surface being lighted (the amount lost is the square of the distance to that surface). At an angle (light aimed to the side), the loss is greater. Thus aimed straight (0° or nadir), the sources closest to the surface will provide the most light and sources further away will yield less. (See Chapter 21.)

from what direction?

The direction light comes from is critical to its effectiveness. For tasks, the direction creates usable or unusable light, reflected and direct glare, and unpleasant facial shadows.

Usable light for reading a book is from the opposite direction of usable light for a computer. For reading, the light should come from behind the person (or from the side). For computers, the light should come from in front of the person (or the side). (The amount of light from the side is often less.)

Unusable light comes from behind a person casting shadows on their visual task. Reflected glare on reading tasks comes from direct light in the 0 to 30 degree range. The reflection obscures (or veils) the printed image, particularly on glossy paper.

Reflected glare can come from any glossy surface (wall paint, glass, polished granite, or brightly lighted surface, etc.). Such glare can be fatiguing to the eye, prevent reading a sign, and botch the make-up job at a mirror.

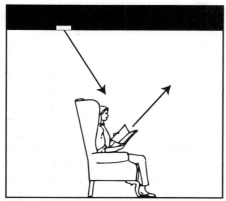
Reflect light away from the eyes.

Direct glare comes from luminaires putting out light in the 60- to 90-degree range (like bare fluorescent strips on the ceiling) or from too much brightness at a window. Direct glare can be fatiguing and distracting. If a window is too bright, more electric light is needed inside to complete visual tasks.

Rule of Thumb for a Bright Window

Too bright a window requires more electric light inside for visual tasks.

Unpleasant facial shadows come from open downlights overhead. Never, never put an open downlight over a mirror or over a chair. Knowing the ultimate effect of the direction of light will enable you to design the best possible lighting.

Light is direct, indirect, direct-indirect, or general diffused. (See Chapter 21.) Direct light from open luminaires provides the greatest amount of light. If incandescent, direct light can create sparkle and enhance colors and texture. Fluorescent does not.

General diffused light rarely provides suitable long-term task light. It can be glaring if bare light sources are too high in wattage, but it also can create a decorative statement if they are chandeliers, pendants, or wall sconces.

Indirect light comes from a lighted surface—ceiling or wall. It provides the greatest spread of light from hidden sources. Indirect wall light can flatter the occupants. A surface for indirect light needs to receive three times as much light as a directly lit surface. The surface needs to be fairly evenly lit from side to side with no hot spots. The light sources for ceiling indirect light must be at least ten inches away from the ceiling.

Never, never have only one distribution of light in a space. It is uninteresting and misses lighting's potential of being the best possible.

Rule of Thumb for Distribution of Light

Always have more than one distribution of light in a space.

---

what sources?

Knowing about the advantages and disadvantages of light sources is the first step in design. New sources are constantly being developed. Energy efficiency of sources is improving. The federal government continually requires higher energy-efficiency standards for light sources. In 1992, the Federal Energy Policy Act (EPACT) outlawed many commonly used sources—like cool-white fluorescent in 40 watts (the workhorse of commercial lighting), incandescent reflectors in 75 and 150 watts, and line-voltage PAR-38 sources in 65-to 150-watt sizes (the workhorse of residential downlights). EPACT limits light sources by the amount of lumens per watt produced—lamp efficacy. It definitely jump-started the light-source manufacturers into developing substitutes—too many substitutes. Now the designer must choose between minimally acceptable, suitable, and best substitutes. Sometimes designers must switch to a different type of source to get the best possible substitute within the constraints of the lighting problem.

Types of Light Sources

Light-source types are incandescent, discharge, laser, electroluminescent, and light-emitting diodes (LED). Incandescent and discharge sources are enclosed in bulbs and are called bulbs by the general public. To the profession, they are lamps and the glass enclosure is the bulb. Since this is confusing, this book refers to them as light sources, which they are. Many sources have hazardous materials inside the glass (particularly the discharge sources) and MUST be disposed of at hazardous waste sites—not put into the trash that goes to the landfill. Teach your clients!

Laser sources are a narrow beam of a single color of light. Electroluminescent sources produce colored light by applying an electrical field to a solid material. Unlike some other sources, their light is not due to heating effects alone. LED's are light emitting diodes. These sources produce light for effect, not for tasks.

INCANDESCENT

Incandescent includes both line-voltage (120) and low-voltage (usually 12 or 24). Low-voltage needs a transformer to reduce power. Halogen is incandescent and can be line- or low-voltage. Halogen describes the gas inside the glass. (Quartz sources describe the glass enclosure.) Some low-voltage sources have xenon gas, which creates less heat.

Nearly all incandescent sources are point sources (except linear incandescent lamps). Point sources can easily be aimed and the beam shaped; linear sources cannot.

All incandescent sources burn hot and much of the energy consumed creates heat—unwanted in hot climates; welcomed in cold climates. Low-voltage sources control the light better for accent lighting. However, high wattage (75 watt) low-voltage consumes as much as line-voltage sources.

COMPARISON OF PERFORMANCE

SOURCE	WATTS	fc @ 6.5ft
line voltage		
PAR/flood	75	55
A	75	13
low voltage		
MR-16/flood	75	89
Q/T4	75	32

Determine how much light can be anticipated from both line- and low-voltage sources at the same wattage before making choices. Luminaire manufacturers' literature supplies this information. Watts are a measure of how much electricity a light source consumes. Use watts when comparing energy efficiency. Footcandles are the amount of light falling on a surface. Use footcandles when comparing amount of light received. Watts cannot be translated into footcandles. In addition, determine dimming capabilities, if needed and expected life of the source.

COMPARE FLUORESCENT SOURCES

4 ft LAMP	LUMENS
rapid start T-8	2,650
rapid start T-12	2,570
high output	3,465
very high output	4,080

Line-voltage incandescents are the least expensive, are easiest to dim, produce shadows and highlights, and are good for accenting and for visible sources in decorative shapes. However, they burn out quickly, consume more energy, and are not good for general room lighting—they create ceiling acne (many holes in the ceiling).

Low-voltage incandescents are easy to wire for remodeling, have many shapes and amounts of light, last longer, can have a remote transformer if not a track fixture, have a whiter incandescent light, and are good for accenting. However, they are more expensive, are not good for high-ceiling mounts, can shatter if fingers handle bare light-producing area of bulb, and are not good for general room lighting. Also, continuous dimming temporarily darkens the bulb.

DISCHARGE

Discharge sources include fluorescent, cold cathode, neon, sodium, mercury, metal-halide, induction, and fusion.

Fluorescents (hot cathode) are made in linear, circular, or compact shapes. Become familiar with the wide array! Fluorescent has many colors, many more than incandescent. Fluorescent is classified by starting characteristics, either needing (pre-heat) or not needing (instant-start and rapid-start) a starter. Fluorescent sources need a ballast to operate. Fluorescent ballasts are either magnetic (the cheapest, but most energy inefficient) or electronic. Ballast type must be compatible with light-source type.

Diameters of linear fluorescent range from 1½ in. (T-12) to ¼ in. (subminiature). A linear fluorescent source can be tucked almost anywhere.

With most linear sources, the length is tied to the wattage. Thus, higher wattage is longer. But high-output and very high-output sources, and some long compact fluorescent sources are the exceptions. They can be the same length but have more lumens. Lumens are the total amount of light a source produces. Use it to compare amount of light available to be received. The in-

EQUIVALENT COMPACT FLUORESCENT SOURCE

INCANDESCENT WATTS	COMPACT FLUORESCENT				
	# TUBES	WATTS	BALLAST	ADVANTAGE	USE
60	2	13	magnetic	small size	sconces, under cabinet, portable fixture
100	4	28	electronic	less electricity, cooler, flicker-free	downlights, wall wash, cove, cornice
150	3	32	electronic	less electricity, cooler, flicker-free	downlights, wall wash
200	2	50	electronic	less electricity, cooler, flicker-free	1 x 2, 2 x 2, recessed, cove, canopy, under cabinet

formation is available in manufacturers' lamp catalogs. Use the mean lumens, if given, because initial lumens do not last; mean is a better measure.

Compacts are bent linear tubes. Some have built-in ballasts; some are encased in glass, like a PAR source. Some have screw bases; others have pins. Never use a screw-based compact fluorescent in an incandescent recessed fixture. The incandescent housing will not redirect the light correctly; specially designed fluorescent reflectors are required!

All fluorescent sources burn cool, last a long time (typically 4 years in a residence), are energy efficient, and are good for large surface lighting. However, they require special dimmers and cannot be completely dimmed, cost more initially, can flicker if used with a magnetic ballast (perceptible to some; not to others), and can create minimal shadow definition of objects if the only source. Very bland.

Cold-cathode is the cold version of fluorescent and is linear shaped. It has great architectural adaptability. It can be hidden or visible, can be made in any custom color and configuration, and can be uniformly dimmed.

Neon is also linear requiring high voltage to excite the gas and make the colors. It is custom made and is usually surface mounted—not a task light.

Mercury, metal-halide, and sodium are called high-intensity discharge describing the pressure. Mercury and metal-halide are suitable for outdoor use, and metal-halide for commercial indoor use. Metal-halide comes in deluxe versions that enhance colors on packaging in big-box retail or supermarket stores. White sodium can be used in commercial and residential spaces

if not the only source type, because of a lag in restriking after being turned off. Sodium fades colors of objects the least and can be low wattage.

Induction and fusion discharge sources have long life and are used for inaccessible, high-ceiling places. They have entirely different source shapes. One induction source is a tube shaped in an open rectangle. Another induction source is a globe bulb. The fusion source is a delicate-looking thin tube with a small glass globe on top. It cannot be looked at directly when lighted without damaging the eyes, since it is created by microwaves.

LASER

Laser sources require a controller and a projector, producing a very narrow, single-color beam made visible by fogging the air. Laser beams are classified by their dangerousness—Class I is considered not dangerous and does not need a permit. All others do. (Lasers are a powerful tool for medical practices.) Lasers produce exquisite light for creating art with light, but do not create task visibility.

ELECTROLUMINESCENT

Electroluminescent lighting in one form is a laminate strip containing phosphors that glow when electricity passes through. It requires a transformer for correcting the voltage. Strips can be up to 1,500 feet long and up to 22 inches wide. The light is brighter than a computer monitor and is offered in several colors.

LIGHT EMITTING DIODES

Light emitting diodes (LED) create visible colors

of light. They can be delightful, colored accent light. They consume very little energy. They also can be used in exit signs, requiring 1.5 watts and lasting for 25 years, clearly, the best choice.

Color of Light

Knowing about the color of light is important. The color of a surface cannot be seen unless the color of that surface is in the color of light. For example, a red apple cannot be seen in blue light. Color of light is controlled by the spectrum of the light source, represented by a spectral distribution curve. Looking at such a curve source does not help the designer know what will actually be seen. Hence, illumination scientists developed two scales to help predict the color seen. The Kelvin Temperature scale (K) indicates the coolness or warmness of light and the Color Rendering Index (CRI) indicates how the source renders surface colors. The scales are not perfect. Use them together and with caution.

KELVIN TEMPERATURE SCALE

Since color of light is hard to describe, Kelvin temperatures (K) indicate coolness or warmness of the light. However, Kelvin temperature is only accurate when referring to sources that can be heated from cold black through red hot, white hot, and on to brilliant blue hot. Only candles, sunshine, or incandescent light comply. Notwithstanding, Kelvin temperature is also used to describe the apparent color of light from discharge sources and fusion. The Kelvin temperature scale ranges from 1,900 to 26,000 degrees, often shortened to 19K and 260K.

With the Kelvin scale, the lower the temperature, the warmer the color of light. The higher the temperature, the cooler the color. The best way to remember the scale is to think of an outdoor scene at sunset with the sun low on the horizon—the sky is red on the horizon (low in Kelvin degrees); the sky high above is blue (high in Kelvin degrees). The memory device is: low temperature—red low in the sky; high tempera-ture—blue high in the sky.

Use Kelvin temperatures to predict how cool or how warm the light will appear. If the space is to be dimly lit, use warm light. If the space is to be brightly lit, use cool light, particularly if the interior colors are to be warm. Do not use overly cool light with totally cool interior colors, nor very warm light with totally warm interior colors. Balance interior color schemes with the opposite temperature in sources.

Never mix different Kelvin temperature sources in a fixture, such as 41K and 30K in the same direct-light ceiling luminaire. The effect is horrid when looking at it, even though the effect is unnoticed when looking at the lighted surface. Colors of light blend on surfaces, but not at the visible source.

The old standby, incandescent, is between 27K and 29K and satisfies most everyone. Consequently, if a new source, like deluxe high-pressure sodium, is announced at 27K, the temperature is a lot better than high-pressure sodium at 21K.

In addition, do not be fooled that the same Kelvin degrees in two different sources produce the same results. If the sources are different types, their ability to render colors may be very different. Consequently, the designer needs to know what the Color Rendering Index is.

COLOR RENDERING INDEX

The Color Rendering Index (CRI) predicts how well colors look when illuminated by a particular source. The highest rating is 100; the lowest is 20. For example, T-8 U-bent fluorescent sources have a CRI of 85. They are available in 30, 35, or 41K versions. But each renders colors differently because their Kelvin degrees are different. Compare CRI only if the sources are within 300 degrees Kelvin of each other. Both Kelvin temperature and CRI must be used together and used cautiously. The comparisons are not absolutes and are based upon subjective impressions of what appears normal.

APPARENT TINT GIVEN TO SURFACE COLORS						
SOURCE	°K	CRI	WHITE	BLUE	GREEN	RED
41K fluorescent	4,100	70-90	none	intense	slight blue	slight orange
35K fluorescent	3,500	70-80	none	crisp	brightens	bright
30K fluorescent	3,000	70-90	light yellow	warms	lightens	slight orange
27K fluorescent	2,700	70-80	light yellow	warms	darkens	enhances
100W incandescent	2,870	100	light orange	dulls	darkens	enhances

DESIGNING WITH COLOR OF LIGHT

Matching all Kelvins and CRIs in one space would be wonderful, but almost impossible when using different sources. The best strategy is to have subtle differences between sources—abrupt differences are unpleasant. Typically, do not have color temperatures differ more than 10K. For example, do not use 41K and 30K fluorescent lamps in the same space.

Disguise differences by shielding the light source from direct view. Also, avoid illuminating parts of the same surface with different sources. For example, do not wash a wall with 41K fluorescent and accent art that spills 28K light onto the same wall from an A-lamp track luminaire.

Since 4-foot, cool-white and warm-white fluorescents are outlawed by EPACT, the choices are 30, 35 or 41K. Use 41K—a cool light—for work environments with light levels above 100 fc (footcandles). Use 30K—a warm light—for living environments (residences, hotels, eating places, and other relaxed atmospheres) with low light levels. And use 35K—a warm but crisp light—for offices, retail, and other spaces. A suitable mix is 35K and MR-16's.

Color of light can give a tint to surface colors—particularly white. Thus, be aware that 30 and 27K fluorescent and line-voltage incandescent give a yellow to peachy tint to white; 41 and 35K fluorescent do not. Thus, for a totally white kitchen, use 41 or 35K sources. Much information goes into designing the best possible lighting.

energy efficiency

About half of the electricity used in commercial buildings is for lighting. Therefore, several states, a few cities, and the federal government have legislated energy efficiency. State legislation limits the amount of watts per square foot. Federal legislation limits the usable light sources determined by lumens per watt. Both intend to reduce energy consumed and to improve appearance, visual comfort, and perceived quality. The legislation does not intend to foster delamping, underlighting, or other methods that reduce light for tasks and affect productivity and comfort. The legislation requires knowledgeable lighting design using task-ambient schemes, the most efficient lamps and luminaires, and the most effective lighting controls. Thus, lighting a workspace wall to wall at the same high level required for the drafting-table task is not appropriate. Therefore, knowing the low-wattage, high lumen-producing sources and efficient luminaires is critical to energy-efficient success. And knowing control-technology efficacy can also increase energy efficiency.

Reprinted by permission of King Features Syndicate, Inc.

what luminaires?

Luminaires can be visible or not so visible. They differ in their capabilities to delivering lumens to the lighted surfaces. Check manufacturers' technical data, choosing the best possible.

Visible fixtures can be hung from the ceiling or mounted on the wall and can make a decorative statement or be subtle. If decorative, they can enhance the style of the space—they are like jewelry. They can also be fiber optics and low-voltage strings or strips. Choices for visible fixtures are almost infinite.

Not-so-visible fixtures can be ready-made (recessed) or custom-constructed (such as coves). Recessed fixtures are engineered to be covered with ceiling insulation (IC, insulated ceiling) or not (TC, thermal cutout). Do not violate their intended use; fire can be the result. Some IC fixtures are also airtight. In areas where radon is in the soil, airtight luminaires save lives by reducing the potential of breathing radon. Use them. Radon is considered the second leading cause of lung cancer. Airtight also reduces heat and cooling loss, and blocks noise intrusion.

Recessed fixtures can put light straight down, to the side, or frame it. They can have diffusers, lenses, louvers, pinholes, cones, or baffles at the opening (aperture). Their appearance does not necessarily indicate how they deliver the light —check the manufacturers' technical data. High quality diffusers and lenses can mask the light source and distribute the light well. Pinholes can allow a broad or a narrow beam of light, depending upon where the light source is positioned. Parabolic louvers are designed to put light straight down (onto a desk surface). Do not use them at a bathroom mirror. They do not light the face well. Use egg-crate louvers that allow light to spread. Some cones reflect the light source and create a glare. Some luminaires are so shallow that the light source sticks out the aperture, creating glare when lighted. Baffles reduce the aperture brightness and are either black or white, depending upon the "look" wanted when off. Some luminaires have sloped ceiling adapters.

COMPARE RECESSED LUMINAIRES

SOURCE	WATTS	fc @ 6.5'	BEAM WIDTH	LIFE
line voltage				
PAR/flood	75	55	4.5'	2,500
A	75	13	8.5'	750
fluorescent	32	13	7.5'	10,000
low voltage				
MR-16/flood	75	89	3'	4,000
Q/T4	75	32	4'	2,000

Compare recessed-luminaire performance in manufacturers' technical catalogs. Check watts, footcandle output, beam width, and source life to select the best possible lighting.

Fluorescent and some 12-volt source recessed fixtures can have emergency backup. Use them even in residences, not just in commercial spaces, where required by code. Protect your clients with light; they could be grateful.

Custom-constructed fixtures can be lighting in arch revels, brackets, cornices, coves, dropped soffits, and perimeter coffers, using either fluorescent or remote-source lighting (single light source remotely mounted with a fiber-optic or a prismatic-film conveyer). Custom-constructed fixtures fit the architecture and can provide excellent general and task lighting. Use them as often as possible.

what lighting scheme?

Schemes for residential and commercial spaces providing services can be task/ambient. Task/ambient schemes supply higher illuminance for visual task surfaces and lower levels for the ambient (nontask) surfaces. Lighting designers will need to use point-to-point calculation methods or manufacturers' technical photometric information for determining task lighting, and use the zonal cavity method for determining the ambient light. (See Chapter 21.)

In addition, brightness ratios must be considered. For long-term visual tasks, surroundings within arm's reach should not be lighted greater than the task amount and not lower than one-third of the task. Surroundings beyond arm's reach should not be greater than five times as bright as the task and not less than one-fifth of the task. Thus, for long-term reading, do not read in a dark room.

Schemes for retail, religious, restaurant, and other special-environment spaces are vastly different. The tone, intent, and visual impacts must be considered to create the appropriate interior atmosphere to be the best possible lighting.

In all schemes, ceiling design can be an excellent platform for lighting, giving organization to the lighting fixtures. Architects and interior designers have opportunities to initially design a special ceiling to integrate lighting. Lighting designers typically do not. If the opportunity is available, all professionals should make use of it!

Professionals who design lighting need to know how to provide visibility for the clients' tasks and to create a pleasing, colorful, and flattering visual environment. With such knowledge, the best possible lighting can be designed.

chandeliers
and
other visible
fixtures

3

Do you have a chandelier hanging over your dining table? Do you think that the room is therefore lighted? Most of it is not. Unfortunately, most dining room chandeliers are equipped to overpower and therefore produce poor-quality light. Usually they are equipped with over-bright bulbs. Sometimes they are wired through a dimmer, but when more light is required chandeliers are turned on fully. At that moment the bright, glaring light comes from the center of the room and tries to light everything. Anyone turning their back to the chandelier creates his or her own shadow. Anyone trying to look across the room is doing so through a blaze of light. A dimmer can subdue the blaze, but shadows and darkness are created instead. Shades on the chandelier can subdue the glare, but they also subdue the only source of light. One source is not enough. Chandeliers must be balanced by other sources to be comfortable and flattering.

Chandeliers and other visible fixtures—pendants and wall fixtures—are like pieces of jewelry, attractive but not necessarily functional. Even though they may be expensive, they are not intended to light a whole space, unless the space is small or used for a short time. Like any jewelry, they are flattering accessories. They are highly visible when lighted, so they should not be glaring. They should supply only soft and glittering light.

Adding other light sources in the same room as a chandelier will not necessarily increase the utility bill. First, all of the sources do not have to be used at the same time. Second, the total wattage now used in the visible fixture can be divided and used in several places around the room. The new total wattage may actually be less than the previous wattage. For example, if a chandelier has five light bulbs of 60 watts each, the total wattage will be 300, and the light will be harsh. Create a new total by:

- Reducing the wattage in the chandelier to 15 for each bulb, for a total of 75.
- Adding three 40-watt fluorescent fixtures behind a valance board, putting light up and down, over the draperies along a 12-ft (36.5-m) window wall, for a total of 120 watts.
- Adding two 30-watt reflector downlights over the buffet on the opposite wall, for a total of 60.

The new total is 255 watts, 45 watts less than the old total. Electricity is saved each time all the fixtures are on and the light is soft, sparkling, and well balanced. Often, the fixture can be used alone or in some combination. Seven possible combinations are:

- The chandelier alone, for setting the table (75 watts).
- The chandelier and the valance light, when looking for the serving platter (195 watts).

A chandelier is like a piece of jewelry.

- The valance light, the downlights, and the chandelier, for a gala dinner party (255 watts).
- The valance light and downlights alone, when dusting the furniture and drapes (180 watts).
- The valance light alone, for brightening up the space on a dull day (120 watts).
- The downlights and the chandelier, for a cozy dinner for two (135 watts).
- The downlights alone, for a nightlight when no one is in the space but you do not want a dark room (60 watts).

If there is a chandelier only, the space can be lighted only one way—by having the chandelier on or off. Why not save electricity and gain more light by using a combination of light sources? These can include wall fixtures, wall-washers, accent lights, downlights, cove lights, and lamps. Not all are on display in lighting showrooms. Most showrooms are a sea of visible fixtures, but other kinds of lighting are also available though not as often displayed. Check the index of this book for the types of light sources available, and ask lighting consultants about them. If you have or want visible fixtures, they can be beautiful, flattering, and energy-efficient jewelry. Learn how and where to use them.

Most visible fixtures need to be electrified.

They look best with the least hardware showing, hooked up directly into an electric junction box. This connection requires prewiring or new wiring. Otherwise, a cord and plug or electric track components are required. Owners, with the help of a good electrician, whether they are building or relighting (installing new lighting), can usually connect directly to the electricity. Renters usually cannot. They must rely on connections with a cord and plug.

Most visible fixtures require incandescent light bulbs. Such bulbs are available in a wide range of wattages, often specified by the manufacturer. However, the specified wattage is usually the highest and not necessarily the best to use. To create soft, flattering light, equip your fixture according to the following rules of thumb, based on whether or not the bulb can be seen.

Rules of Thumb for Incandescent Light Bulb Wattage

- For visible light bulbs, use up to 15 watts.
- For obscured but somewhat visible bulbs, use up to 25 watts.
- For hidden bulbs, use 40 watts maximum or the manufacturer's specifications.

Try the lowest wattages first; often the lowest one provides sufficient light. Your visible fixture should enhance the space, not overpower it.

A fluorescent screw-base bulb can substitute for an incandescent bulb in a fixture that hides the bulb, as long as the required amount of illumination is provided. Some pendants can accept an adapted-circle fluorescent tube or a screw-base fluorescent bulb, if it can be well hidden. Use any size that fits and gives you as much light as you want. Some visible fixtures are made specifically for fluorescent sources; use the one specified by the manufacturer.

Visible fixtures have decorative and human impact. They must relate to the scale of the room—to its length, width, and height. In general, avoid large fixtures in small spaces and small fixtures in large spaces. The decorative impacts of visible fixtures are the mass of the fixture, the intended impact, the proportions, and the harmony of the whole space. The human impacts are the visual comfort, the amount of light needed, and the unthreatening position of the fixture. Consider these impacts carefully and you will never regret where you have hung your lighting jewel.

chandeliers

Where to Use a Chandelier

Historically, chandeliers were hung in the center of a room, well above people's heads, to light as much of the room as possible. Chandeliers can still be used in the center of a room in period style and in some contemporary rooms for special effects. In addition, chandeliers can be used above all sorts of tables—dining tables, end tables, coffee tables, corner tables, and game tables. Likewise, a chandelier can be used above a desk, a bar, or any piece of furniture that is not a seat. A chandelier above a seat, hung at any height, can be threatening, making people uneasy. Linear chandeliers are called library or billiard lights. These chandeliers accentuate the geometry of any rectangular table. Over desks and library tables, they provide light to the right as well as to the left. Over billiard tables and dining tables, they illuminate the activity for the length of the table.

A chandelier can be positioned off-center above a buffet, dresser, or vanity top. In fact, small dining rooms appear larger when a chandelier is moved from the center of the table and placed off-center above a buffet against the wall. Small bathrooms are enhanced with a small chandelier to the left or right of a sink in front of a large mirror, reflecting the sparkle.

A contemporary ceiling fixture becomes a chandelier when centered over an eating table. Do not neglect these fixtures as a possibility for an effective piece of lighting jewelry.

Finally, a chandelier hung so that it can be seen through a window creates a pleasing architectural surprise, both inside and out. When the glimpse of it is momentary and later revealed fully, the visual excitement is heightened. Chandeliers hung in unusual places create a tasteful visual surprise. The possibilities of places for chandeliers are almost endless and probably limited only by one's pocketbook.

Linear chandeliers are called library or billiard lights.

A small chandelier in front of a large mirror enhances a small bathroom.

A contemporary ceiling fixture gives a chandelierlike effect.

A chandelier seen through a window creates a pleasing architectural surprise.

How to Choose a Chandelier

A chandelier can be of the same style as the furnishing in a room, thereby carrying out the harmony. For example, a wrought-iron chandelier is congruent with Early American, rustic, or contemporary furnishings. An Oriental faux-bamboo chandelier blends with Queen Anne, Georgian, or neoclassical furnishings. The possibilities of harmony are numerous. A chandelier can also be of a different style, thereby carrying out a counterpoint to the style of the room. Antique classic furniture can be combined with a gleaming chrome fixture, contemporary in style. Smooth Eero Saarinen furniture can be combined with a chandelier dripping with crystal. Create harmony or counterpoint; mix or match.

Size of a Chandelier

Hung in a room and not over a table, a chandelier must relate to the size and scale of the space. Rooms with high ceilings create a large volume and accept larger chandeliers than smaller rooms. High ceilings are enhanced with two-tiered chandeliers, which would be unsuitable in average-height spaces.

Rule of Thumb for a Chandelier Not Hung Over a Table

The chandelier can be in inches what the diagonal of the room is in feet. (For example, if a room is 16 ft on the diagonal, the chandelier could be approximately 16 in. in diameter.)

In the metric system the rule is that the chandelier's diameter can be in centimeters what the diagonal of the room is in meters divided by 0.12. (For example, a room 16 feet on the diagonal is 4.88 meters; 4.88 divided by 0.12 is 41, and the chandelier would be in proportion around 41 centimeters.)

Two-tired chandeliers enhance high-ceilinged spaces.

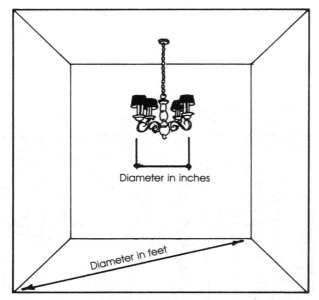

A chandelier can measure across in inches what the diagonal of the room measures in feet.

A chandelier can be 12 in. smaller than the diameter of the table.

This rule about size is merely a guide and can be modified for decorative reasons. If the chandelier is open and delicate, it appears to be smaller and a larger size might be suitable. If the chandelier is bulky and solid, it appears larger and acts as a visual barrier, so a smaller version might be more suitable. In general, big chandeliers need big rooms and small chandeliers need small rooms. The results can be breathtaking. A chandelier hung over a table

Reprinted by permission of Oldden.

must be small enough to allow people to get up and down without bumping into it.

**Rule of Thumb for a Chandelier
Hung Over a Table**
The chandelier should be 12 in. (31 cm) smaller in diameter than the smallest table dimension.

At game tables, where the action is more vigorous, a chandelier should be even smaller. At rectangular tables a linear fixture (library or billiard chandelier) should not exceed two-thirds of the length of the table.

How to Hang a Chandelier

A chandelier hangs from the ceiling by one of three devices: a chain and cord, a cord only, or a stem. Chains and cords can be easily shortened or lengthened. Stems can be shortened, but cannot be lengthened. (Sometimes longer stems can be ordered from the manufacturer.) A canopy cover obscures the place where the hanging device and the ceiling meet. A chandelier is hooked up to the electricity in most cases through an electrical junction box, as required by many building codes. However, a junction box is not always located in the best place on the ceiling. Do not despair. Several alternatives are possible.

New ceiling junction box. Install a new ceiling junction box if electrical wires are accessible, that is, if wires can be pulled from elsewhere inside the ceiling. Access is gained through an attic or by pulling wires between the ceiling joists. If access is available, this is the slickest method. Owners who are building or relighting will want to use this method, even though it usually requires an electrician. An electrician's fee is not prohibitive when compared to the investment in the chandelier itself.

Another ceiling junction box. If another ceiling junction box is available elsewhere on the ceiling, attach the cord there and bring it over to the chandelier by one of the following five methods:

1. Weaving the chandelier cord through a chain and holding it by hooks from the other ceiling junction box.
2. Putting the chandelier cord into a canopy and extending the cord over to the junction box.
3. Knotting and hooking the chandelier cord over to the other junction box.
4. Hiding the wire from the junction box in a raceway channel (a U-shaped metal covering).
5. Connecting the chandelier to a track adapter and fastening it into an electric track wired at the junction box.

All of these connections are highly visible because hardware is put on the ceiling. Of the five, the electric track offers the opportunity to change the position of the chandelier, which is useful to some but wasted by others. The purchaser, of course, pays for this adjustability. However, as the popularity of track lighting has increased, the cost has come down. If it is desired, most chandeliers can be adapted to electric tracks. The limitations are the weight of the chandelier itself and the diameter of the cord or thread size of the stem, which connect to the track adapter. Check the manufacturers' catalogs to coordinate the chandelier's components with the electric track requirements before you purchase; they must be compatible.

Owners who cannot have a new junction box installed and renters who want to move their chandelier will want to use the electric track method of installation. (Consult Chapter 7 about the track clip-adapter method of chandelier installation before making a choice.)

Existing baseboard receptacle. A chandelier can be connected by a chain, cord, and hooks across the ceiling and down the wall to an existing baseboard receptacle. This method is often called swag. Swag is the last type of installation I would suggest, because it detracts from the aesthetics of the fixture; too much hardware shows. However, both owners who have purchased living units in structures that are not suitable for rewiring and renters who want to invest minimally in chandelier installation will

A chandelier should be 2 ft. 6 in. above the table.

want to use this method. An owner planning to build should never use it.

Once the electrical connection is made, the next decision is how high to hang the chandelier.

Rules of Thumb for Hanging a Chandelier

For center-of-the-room installation in average-height rooms, hang the chandelier close to the ceiling. In high-volume spaces, hang the chandelier lower so that it is still in view.

Above an eating table, the lowest part of the fixture should be no less than 2 ft 6 in. (0.76 meters) above the tabletop.

Off-centered over desks, sideboards, bars, and stairwells, the chandelier should be pretested to make sure that sufficient headroom is provided for people to move about and to use the space and furniture.

A head and a chandelier do not mix—one gets bruised, the other gets broken. Likewise, a chandelier must never be hung in a position that threatens a person—over his or her head, for instance. Human comfort is a lighting goal.

A Downlight in a Chandelier

When it is over tables or table-height surfaces, a chandelier becomes more than just a pretty jewel if it contains a downlight. A downlight provides direct, bright light onto the surface below. It can be a working light for short-term activity, such as writing a letter or reading the mail. In addition, the direct light will produce sparkle on tabletop objects: glasses, dishes, and silverware. If the chandelier is to be used over an eat-

Chandelier light.

Downlight.

Downlight and chandelier.

ing table, gain double sparkle—the sparkle of the chandelier itself and the sparkle of downlight on the objects below. Downlight enhances the color and appeal of food, qualities of tableware, and focuses attention to the table. All great advantages!

The disadvantage of downlight is that it can create reflected glare on anything glossy— glasstopped tables and/or glossy reading materials. Therefore, do not use a downlight over a glasstop table! Do not use a chandelier or pendant with direct sources pointing down! Do not use chandeliers with glass diffusers hanging below! The reflections are unrelenting. Choose a chande-

lier with sources pointed up. (Observe how often in magazine photographs, a large bowl of flowers is on a glasstop table to obscure the glare of a downlight.)

Rule of Thumb for Downlight over Glasstop Table
Never use downlight over a glasstop table. It creates unrelenting glare.

In addition, open sources pointing down from pendants or chandeliers create direct glare in the eyes of the people seated at the table—visually distracting.

Downlights typically accept a 50-watt R-20 reflector. Downlights could have more punch with a narrow spot in either a 45-watt halogen PAR-16 or a 50-watt PAR-20.

Manufacturers of chandeliers typically specify 60 to 100 watts maximum wattage allowable per arm of the chandelier. Thus, a six-arm chandelier could have 600 watts total—a glare bomb! Pendants typically have 100- or 150-watt A or G standard sources specified, which are energy-inefficient. Reduce the wattage to 25, 40, or 60 with A- or G-halogen and get three times the life.

Glare can even be created without downlights. Bare-light source chandeliers can reflect glare on other glass surfaces—a mirror or an undraped window at night. If not overpoweringly bright, the reflection can be a glimmering bonus— two visual images instead of one. Brilliance should come from the decorative effect of the chandelier, not the wattage of the light sources.

wall fixtures

Wall fixtures illuminate from the wall. They come in numerous styles:

- Sconce (half of a chandelier).
- Track fixtures on an electric track.
- Wall-hung lantern.
- Bar-shaped fixture with bulbs behind.
- Accent lights, pointing in one direction.

Where to Use Wall Fixtures

Use wall fixtures in a dining room above a buffet, on a large mirror, up the steps, gracing an arched doorway, beside the bed, or in other hard-to-light places. Furniture is not needed below them.

Use wall fixtures to illuminate semipublic business spaces where a less officelike appearance is desirable. Restaurants, shops, restrooms, waiting rooms, and professional offices can appear warmer and more inviting with appropriately styled wall-hung fixtures.

Wall fixtures can be used alone or in pairs. They can match a chandelier or another fixture in the room. They can be centered, off-centered, or used in pairs. However they are used, they can accessorize a space by themselves.

How to Choose a Wall Fixture

Sometimes you choose a wall fixture to enhance the style of the room or structure. If so, always choose a fixture that matches another visible fixture. For example, a pair of outdoor wall lanterns hung beside a Federal mirror above a dressing table match a lantern outside on a balcony. If there is no other fixture to match, choose one in a different style to enhance the interior design. It can be successful if carefully planned.

Some wall fixtures are adjustable. Some

Match a wall fixture with a chandelier.

A pair of outdoor wall lanterns hung beside a mirror look effective with a matching lantern on the outside balcony.

are stationary. Both provide light in difficult places. Use adjustable wall fixtures to light a desk, a bed, or an easy chair where there is no space for a table lamp. A track fixture and track canopy can function together as an adjustable wall fixture.

Some wall fixtures allow the bare bulb to be seen and should provide only a soft amount of light. Some fixtures direct the light above and below. Some point the light in one direction to accent objects. They are difficult to use successfully and need other light sources for balance and softening.

Sometimes a wall fixture can create a sense of scale. The size of the fixture must be compatible with the wall's height, the furniture, and the volume of the space. To determine the size, test it.

See for Yourself: Is the Wall Fixture the Right Size?
1. Build a rough model with cardboard and masking tape as close as possible to the shape and size of your wall fixture. (Shoe boxes are often the size of a wall fixture.)
2. Hold the model up to the wall.
3. You can not only judge the size but also decide how high to hang it.

How to Hang a Wall Fixture

There are two ways to connect a wall fixture to the electricity. Connect the wires to the electricity in a junction box behind the fixture in the wall. This installation is called wall mount or outlet-box mount. Owners who are building can obtain a junction box by having the wires installed originally, or owners who are relighting can pull the wires from a baseboard receptacle below the desired location of the fixture. Skilled electricians can pull wires from seemingly impossible places.

Using the wall mount or outlet-box mount to connect a wall fixture is the most aesthetic method because the cord does not hang down. The cord is objectionable to many people. Over and over again, the readers of my newspaper column on lighting expressed their displeasure at seeing cords.

If there is no junction box, plug the electric cord into a baseboard receptacle. This installation method is called pin-up, or cord and plug. When purchasing a fixture, specify which type of installation is to be used.

Renters must use the cord and plug method. Sometimes owners choose to do so. Be sure that the electric wire is long enough to reach the nearest baseboard receptacle. Otherwise, have the pin-up rewired. Some have hollow tubing or a chain to obscure the cord. The tubing comes in 1-, 2-, and 3-ft lengths (0.3, 0.5, 0.8 m). Check the manufacturer's catalog to see if such accessories are available for the fixture chosen.

Rule of Thumb for Hanging Wall Fixtures
All wall fixtures are hung at or near eye level.

Wall fixtures are meant to be seen. Hang them higher only if they infringe on the space needed to move around; for example, beside steps.

Uplight and Downlight in Wall Fixtures

Wall fixtures can be uplights and/or downlights. Be careful that no one is blinded by the glare or dazzled by the light, especially on the stairs, under balconies, in upper bunks, and in other up and down places.

See for Yourself: Is There a Glare?
1. Hold a flashlight where the wall light will be.
2. Have someone check the glare above or below.

Alternative Energy Sources for Wall Fixtures

Frequently, readers of my newspaper column would ask about battery-operated wall fixtures for their living spaces. They are available for utility spaces where electrical connections are difficult. They are, however, very utilitarian looking. For more formal spaces, wall fixtures using other energy sources are better.

Manufacturers make candle- and kerosene-fueled wall fixtures. Some are elegantly styled. The light produced by these fixtures is pleasing, comfortable, and never glaring. But it is produced by fire. Consequently, the fixtures require monitoring as well as cleaning. Notwithstanding these inconveniences, give these fixtures more than just a passing thought in your deliberations. Candle or kerosene light creates a festive mood.

pendants

Pendants are ceiling-hung fixtures similar to chandeliers but usually without branches. In some cases pendants spread the light. In other cases they push light down like a downlight. Transparent or translucent pendants spread light; opaque pendants push it down. If light needs to be distributed broadly in the space, choose a transparent or translucent pendant. If a worklight or a concentrated pool of light is wanted, choose an opaque pendant. These often create cozy or intimate light.

Hanging devices for pendants are like the ones for chandeliers—cords and chains, cords, or stems. Some cords have pulleys that make the pendant adjustable up and down. (The British aptly call pulleys rise-and-fall mechanisms.) A pulley is especially good for those who cannot or should not climb up to change a light bulb.

Where to Use a Pendant

A pendant can be used in the same places as a chandelier: centered above a table and off-center above other furniture except a seat. Over a seat, a pendant becomes a threat to one's head. Pendants appear natural over other furniture, and they also emphasize architecture without furniture, such as a dramatic open stairway, a high ceiling, or a massive fireplace. They call attention not by lighting the architecture but by being associated with it. They fit in small spaces, such as a narrow window. They can light an area that needs bright, direct worklight. Pendants can be clustered from a common canopy or hung in rows independently. They can be a repeatable design element.

Opaque pendants can create an intimate mood if hung at eye level, casting their light down at an eating table, at each end of a sofa, beside a bed or lounge chair—anywhere intimacy is intended. In professional offices, pendants make the atmosphere less institutional. Like chandeliers, they can be either very expensive or reasonably priced. Good judgment and good taste will help you choose a reasonably priced pendant that looks expensive.

How to Choose a Pendant

Pendants come in various shapes—domes, drops, lanterns, shades, and globes. Some pen-

A pendant can provide a repeatable design element.

dants emulate the styles of yesteryear, and some are in styles not associated with any historical period. Mix or match them with your interior design. Beyond style, pendants can emphasize activities. For example, a stained-glass pendant containing the Queen of Hearts, Jack of Diamonds, King of Clubs, and Ace of Spades over a game table reinforces the intended activity. Use what is appropriate for the space.

A pendant must relate to the size and scale of a space. A high-ceiling space can accept a larger pendant than a low-ceiling space. A small space must not be overwhelmed with a big pendant without first carefully considering the decorative effects. This is not to say that a large pendant should never be used; it can be, if it is carefully choosen. For example, a large, paper-covered Japanese lantern is appropriate in some small spaces. But, before buying, make a rough model in the general shape of the pendant being considered, hold it up, and judge it in the space.

Pendant emphasizes an activity.

The rule of thumb for the size of the chandelier given earlier in this chapter is also a guide for a pendant, although opaque pendants need to be smaller, usually by about 20 percent. If it is undersized, however, a pendant looks puny. Both too big or too small pendants are unsuitable. A lighting fixture should never stand out because of its size, whether it is large or small.

How to Hang a Pendant

Hang a pendant from the ceiling by hooking it up to the electricity through a ceiling junction box. If a ceiling installation is not possible, it can be hung from the wall as a swag. But a pendant looks best when connected directly through the ceiling. Another choice of hanging is to adapt it to an electric track. A track can reach from an existing ceiling junction box located in an undesirable location. A pendant on a track can be movable.

When hanging a pendant over a table, follow the rules of thumb for hanging a chandelier given earlier in this chapter, or follow these rules:

Rules of Thumb for Hanging a Pendant

- Alongside a chair or sofa, the bottom of the pendant should be about 42 to 45 in. (a little more than 1 m) from the floor.
- Beside a bed, the pendant should be 24 in. (61 cm) from the top of the mattress.
- Above a desk, the pendant should be about 20 in. (51 cm) from the surface.
- Not over furniture, the pendant should be hung at eye level.

Light and Color from Pendants

Pendants hung at eye level are governed by the rules of thumb for incandescent light-bulb wattage: Visible light bulbs must be 15 watts or less; somewhat obscured bulbs can be up to 25 watts; and hidden bulbs can be as high as 40 watts unless otherwise specified by the manufacturer. If a reflector bulb is specified for your pendant, use it. A reflector puts out a greater punch of light from a smaller wattage. Even if the manufacturer says the highest reflector wattage usable is 75, try 50 or even 30 watts. Lower wattage might produce enough light, and it will save electricity. Pendants hung above eye level can have higher wattage light bulbs, but be careful not to overpower the space with glare. Higher wattages need to be softened with other light sources.

If you want to get as much light as possible, or you prefer an active mood in the space, choose a pendant that allows the light to go up as well as down. Light will be reflected from the ceiling. If you want to achieve a more passive mood, use a downlight pendant. Light will be more concentrated and can be subdued. Make sure it is not glaring. Some downlight pendants have scientifically designed louvers to cut off the view of the bulb and reduce the glare from the bottom of the fixture. They are very good for seating areas. Always choose a pendant with a louver or lens if it can be seen from below. If the pendant is viewed from above, choose one that blocks the light. Comfort is worth considering when choosing and hanging a pendant.

Translucent pendants made of colored materials—glass or plastic—can cast a color with the light. Be sure the color does not create unpleasant effects. In addition, carefully coordinate the color of a pendant with the background against which it will be viewed. To blend with the least contrast, a pendant should be the same color as the background. To create a pleasing contrast, pretest a pendant of contrasting color against its background—if not with the actual fixture, at least with a piece of similarly colored construction paper. In addition, make sure the pendant's color is repeated elsewhere in the space so that it is not an intruder. If it is carefully chosen, a pendant can be an accent, not an accident.

selecting the light source
style for visible fixtures

Many different types of light sources are available for visible fixtures, particularly for bare light-source decorative fixtures (chandeliers and wall fixtures). If the sources are well chosen, they can flatter the fixture and the space. The list of types is long, but in most places the supply is short. Do not be deterred. They can be obtained.

The glass in incandescent sources can be clear or coated. Clear glass permits seeing the filament. Coated glass obscures the filament. Some clear sources are tinted—iridescent, amber, ruby, smoke, blue, and green. The tint adds a slight hue to the light. Coated sources are red, orange, etc., but white is the most common. Frosted sources are not the same as coated. They have hot spots, because they do not diffuse as well as white coated sources.

If incandescent is the choice, clear light sources should be used when the source is seen, and coated sources should be used when the source is shielded (by a shade or opal globe). However, rules are meant to be broken if a better result is obtainable. Experiment! Do not spoil an attractive bare-source chandelier or wall fixture with an indifferent incandescent light source. There are so many to choose from.

The rule sounds backward, but it is not. Fixtures displaying bare sources are designed to show off the lighted filament as though it were a candle flame. Consequently, if the filament is obscured, the aesthetics of the design are destroyed. However, rules can be broken with good cause. Experiment. But, never create glare with a clear source. Use 15 watts or less!

Rule of Thumb for Wattage of a Visible Incandescent Source
Never go above 15 watts, ever.

Halogen candelabra sources last three times longer than nonhalogen. Also, A-Shaped compact fluorescent last four times longer and also are dimmable withdimmable electronic ballasts. Low-voltage cable, rod, and rail systems, which hang wiring and fixtures in full view, use MR and bi-pin low-voltage sources.

Manufacturers identify light sources by a letter and a number code. The letters describe the shape. Often, the letter stands for a word describing the shape, like CA for candelabra, F flame, G globe, and T tubular. The letter A, however, does not. It is designated as "arbitrary," and is considered "standard." The teardrop, candelabra, and flame shapes (B, CA, and F) are the traditional shapes for bare-source chandeliers and wall fixtures. The globe, sphere, and candle shapes (G, S, and C) are good for stylized fixtures. These non-flame shapes are different, not indifferent. In contemporary fixtures with clear-glass, square-shaped covers, a tubular (T) source repeats the linear style of the enclosure. With clear-glass, round-shaped covers, a globe (G) source repeats the round style of the enclosure.

Clear light source when seen.   Coated source when not.

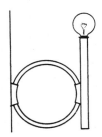 

Line-Voltage Incandescent Shapes.

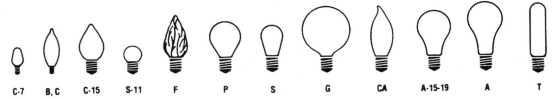

Reprinted by permission of Osram-Sylvania.

screw-base   pin-base   R   G   2D

Compact Fluorescent Shapes.   *Reprinted by permission of Osram-Sylvania and General Electric.*

The United States number code indicates in eighths of an inch the diameter of the glass bulb at the widest point. Thus, an A-17 is 17 eighths of an inch or 2 ⅛ inches (53 mm) wide. The CA-8 is 1 inch (25 mm) wide.

A chandelier or wall fixture is equipped with a socket to accept the size source it was designed for: a line-voltage incandescent with candelabra or a medium-screw base, or a pin- or screw-base compact fluorescent in a variety of shapes. Adapters can convert a medium-base socket to hold a candelabra, or convert a pin socket to a screw base, if necessary. All are line-voltage. Screw sockets accept either incandescent or screw-based fluorescent.

Whenever possible, choose a visible fixture that hides the source and could use a compact fluorescent. It saves electricity and lessens harmful atmospheric emissions. A 9-watt, screw-base fluorescent puts out as much light as a 25-watt incandescent but uses 16 watts less electricity to do it.

Some visible fixtures are specifically designed for compact fluorescent sources. No adapter is made to convert back to incandescent.

Fluorescent wall fixtures can become emergency lights with a remotely mounted emergency ballast. Very useful for residences and residential-like commercial spaces.

Pendants with opaque, translucent, and transparent shades can utilize compact self-ballasted fluorescent sources. A compact source must have a ballast. The electronic self-ballasted compacts switch on with no flicker. They are suitable for hotel guest room, residence, and office applications. Most cannot be dimmed. The magnetic self-ballasted compacts switch on in a few seconds and are for spaces needing constant light—lobbies, hallways, exteriors.

A compact source lasts 10 times longer than an incandescent. Some compact sources have replaceable bulbs reusing the ballast. These are less expensive in the long run.

Some opaque, direct-light pendants can use a reflector compact fluorescent (R), thereby replacing a 75-watt incandescent flood source with a 15-, 18-, or 20-watt compact source. Substantial energy savings will be gained. The color of light will be a little different; they have a different color spectrum than incandescent. Test it!

If the pendant is meant to have a globe-bulb look (G), get a G compact fluorescent. They are available in 15 or 18 watts. They replace 60-watt incandescent sources. The fluorescent sizes are G-30 at 3 ¾ in. (94 mm) and G-40 at 5 in. (125 mm) wide, consume 15 watts, have a color temperature of 28K, and last 9,000 hours.

Sample Electric Cost

Replace a 75-watt flood with a 15-watt R-30 compact fluorescent and save $60 over the life of the source at 10¢ per kilowatt-hour, including lamp replacements.

---

lamps
to light up
your life

4

Lamps can light up your life. In small commercial spaces, lamps are not often used; fluorescent fixtures are. Consequently, lamps are refreshing. In any space they lend a sense of scale and add warmth not possible with ceiling fixtures alone. In residences, on the other hand, lamps are used too often. Lamps have limitations. In many ways, you'll find that other types of lighting described in this book are better for residences.

limitations

Usually the light created by lamps is an accident of placement, not the intent of lighting design. A lamp is limited by the length of its cord and can be positioned only where feasible. Consequently, within a space, light is created in spots here and there. The spots are visible on the furniture, the floor, and the ceiling. The ceiling spot can reflect and spread the light, but it can also be bright and distracting. Spots of light on the furniture and on the floor may be useful, but often they are not.

Lamps are also limited by the characteristics of the shades. Translucent lampshades can deliver light through the shade, but they become bright. Opaque shades cast light down or up in a size dependent upon the width of the shade's opening and in an amount dependent upon the bulb's wattage. The light can be harsh.

Lamps are usually limited to being near or on furniture. Table lamps are restricted to table-height furniture. Floor lamps are ordinarily, but not always, restricted to a chair, table, or group of furniture. The result is that light is not always in the best place.

In addition, lamps cannot systematically light anything. Neither table nor floor lamps, for instance, can wash a wall with light from end to end. Washing a wall with light makes a space appear larger and feel more hospitable. It flatters the occupants, and enhances the furniture. Do you know a lamp that can do that?

Finally, lamps are limited because they are chosen for their looks, rather than for the kind of light they produce. Clearly, different lamps and different shades produce different kinds of light.

benefits

In spite of their limitations, lamps do have benefits. They are so easy to install—just plug them into a baseboard electrical receptacle. They are portable and have instantaneous installation.

Lamps are excellent supplements to fixed lighting—ceiling lights, wall lights, chandeliers, and all built-in fixtures. Also, for many spaces, lamps provide light where no other lighting solution is possible. They can light difficult places—a tiny area or a place where other lighting cannot be installed. For instance, an arc lamp or stretch floor lamp can put light down onto a group of furniture in a steel-and-concrete condominium where a ceiling fixture cannot be used, or it can reach over a sofa to illuminate reading material where the space is not sufficient for a table and a lamp.

Wall receptacle.

Lamps have another function; they give a sense of scale to the space. Most sizable rooms should have one lamp. Choose a lamp in scale with the room and the furniture—big rooms with big furniture require big lamps; small rooms with small furniture need small lamps.

lamp choices

If lamps are selected for their looks only, they are simply a decorative ornament and should not be expected to provide adequate light for the space they light. Like a chandelier, which is also "jewelry," decorative lamps must be backed up with other light sources. If, on the other hand, they are selected to provide light for a specific purpose, they can be functional as well as decorative. Always choose a lamp for the type of light desired—confined and directed or broadly spread. Lamps that confine the light are those with opaque shades, such as metal-shaded desk lamps or torchères. Lamps that broadly spread the light have translucent shades, such as table lamps with off-white silk shades.

Light confined.

Light diffused.

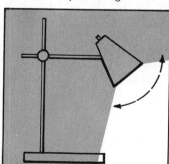

Direct and adjustable light.

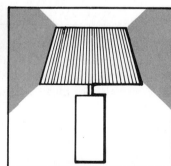

Transluscent shade.

shades

Lamp bases control the height from which the light will be distributed, and lampshades control the distribution of the light. Shades are either translucent (permitting the light to filter through) or opaque (not allowing the light to filter through). Never judge the color of a translucent shade until you see it lighted; it often changes color completely. Be aware of the distribution characteristics of shades.

Rule of Thumb for Distribution of Light by Lamp Shades
The wider the opening of the shade, the wider the light will be spread. The spread of light will be more noticeable with opaque shades than with translucent shades.

If a table beside a seat is to be illuminated, make sure that the shade is wide at the bottom. If a large area is to be illuminated by reflecting light from the ceiling, make sure that the shade is as wide at the top as at the bottom. To ensure that the shade functions as it should, follow the rules of thumb for the bottom of the shade; they vary with the position the shade will be in.

Rules of Thumb for the Bottom of the Shade
A lamp with a shade directly beside a person's head, such as a lamp on an end table, should have the bottom of the shade at eye level.

A lamp with a shade above eye level should be placed behind, at the right or left rear corner of the seating.

A lamp with a shade above a person's head, such as a floor lamp used as a downlight over furniture arranged in a group, should have a louvered or diffuser-covered opening to obscure the light bulb.

The inside surface of the shade affects the amount of light reflected. It should reflect as much as possible. Contrary to what you might think, metallic surfaces—silver and gold—re-

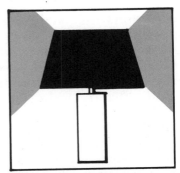

Shade narrow at top and wide at bottom.

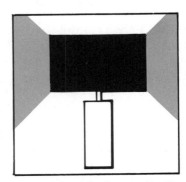

Shade wide at the top and bottom.

flect poorly and can be glaring. Overall, shades should be deep enough to hide the bulb, wide enough to spread the light, dense enough to obscure the lighted bulb, and lined with white to reflect the greatest amount of light.

Lamps themselves are highly visible and produce their light without any mystery about the source. Use the shade to protect the light bulb from being seen from the usual positions in the space—seated and standing. If a mirror is to be illuminated, make sure that the bulb does not glare into someone's eyes and that the light is transmitted to the face. Likewise, if a table lamp beside a chair is desired, do not select it until the height of the table and the chair are known.

light sources

The light source for lamps can be chosen from many wattages and several types, whether they be incandescent and fluorescent. They fit into the lamp sockets, which are either metal or ceramic (resembling white china).

Incandescent

Coated standard A bulbs are good for most lamps. They are manufactured in either single or triple (three-way) wattages. Use three-way bulbs if lamps are the only source for lighting. Determine where to place them for the best effects. Sometimes turn on two or three lamps at the lowest wattage; at other times turn on two at full and one at lower wattage. Control is at your fingertips. Make lamps work for you and save energy at the same time.

<hr>

Sample Electric Cost
A three-way bulb used at 50 watts instead of 150 watts will save 100 watts or $1.20 per month at 10¢ per kilowatt-hour and 4 hours per night. However, the option for 150 is available when needed.

<hr>

Some lamps, usually of better quality, indicate the maximum wattage usable. Do not exceed the indicated wattage. Some lamps do not indicate a maximum. Use a wattage suitable for your purpose. But do not overwatt a lamp so that it produces glaringly bright light; install other light sources. In addition, never install an incandescent reflector or a PAR bulb in a metal socket on a lamp; install it in a ceramic socket only.

Fluorescent

Fluorescent sources last longer, put out more light for the watts, and are cooler than incandescent. Thus, they reduce greenhouse-gas and acid rain-producing emissions, paying for themselves. Need we say more?

Self-ballasted compact fluorescent sources

that fit are suitable. Electronic ballasts turn on without flicker; magnetic do not. Your choice.

If the shade clips onto an incandescent A source, an A-shaped 11- or 15-watt compact fluorescent is needed. If the shade is held by a harp, the compact must fit inside the harp or go around it. Self-ballasted twin-tube, pin-based with an adapter, or a twisted-tube compact (in 32 or 42 watts) fit inside the harp. Those with dimmable electronic ballasts are adaptable.

The 2D or circle fluorescent goes around the harp. The 2D has a three-way version (15-25-39 watts). Use compact fluorescent whenever possible!

EQUIVALENT AMOUNT OF LIGHT

COMPACT FLUORESCENT WATTS	=	INCANDESCENT WATTS
9		25
11		40
13 or 15		60
39		100
42		200

The color of compact fluorescent light varies. Most are 27K, some are 30K, and some pin-based sources are 35 and 41K. Compacts have different spectral distributions than incandescent. Check the effect of the color of light through translucent shades (especially silk). Opaque shades are not affected.

amount of light

The amount of light seen in the space will depend upon the other light sources and the interior finishes. To observe how the light from more than one lamp blends with other lamps, test several lamps as described in Chapter 7. In general, strong directional light from an opaque shade in a dimly lit space will contrast with the surrounding darkness and be irritating. Likewise, an overly bright translucent shade against a dark wall causes too much contrast. Shades that blend with their background are usually considered more pleasing, such as pale-colored shades against pale walls and dark shades against dark walls. However, contrasting shades can be used if they are well planned.

Determine if your translucent lampshade

is too bright when lighted. Follow the directions in the section on measuring light in Chapter 2, but put the meter in the middle of the shade to measure the footcandles.

<hr>

Rules of Thumb for Shade Brightness
- For general illumination in a room with deep-colored walls or wood paneling, the shade should indicate no more than 50 footcandles on the meter.
- For general illumination with pale-colored walls, the shade should indicate no more than 150 footcandles.
- For illumination to perform specific visual activities, the shade should indicate between 250 and 400 footcandles.

<hr>

designing with lamps

The best lamps have an undershade device to soften the light—a diffusing bowl, a refractor, or a disc. However, most lamps do not. Even though undershade devices improve the quality of the light, they are not always necessary if the light in the space is well distributed and blended. Design well-distributed lighting by utilizing the other lighting described in this book as well as lamps. Decide where light is needed for activities, what can be emphasized with the light, and how to balance it.

See for Yourself: Where, What, and How Much?

1. Determine where the electrical baseboard receptacles are and what furniture is appropriate to hold or be related to lamps.
2. Test various bulb wattages in lamps at these locations with both opaque and translucent shades.
3. Decide which positions, shade types, and bulb wattages distribute and blend the light satisfactorily.

Make light perform the way you want it to, and enhance your spaces—at work or at home—for many different moods, effects, and needs. In many ways the vocabulary used to describe how you want the light to look determines how you light it. If it is described simply as "light," you will not have any criteria to make decisions. Describe what you want with many adjectives and adverbs and the solutions will be easy.

Examples

- A dual-purpose light at the desk to evenly illuminate the reading material in various positions and to call attention to the colorful print on the wall above. (The solution is a neutral-colored translucent shade on an adjustable desk lamp.)
- A resplendent soft light at the coffee table, emphasizing the seating group and creating a subspace. (The solution is an arc floor lamp, a lighted coffee table, or, if possible, a metal pendant hanging from the ceiling, which would also create intimacy and emphasize the architecture.)

- A harmonizing, soothing light at the end of the seating group, bringing the dark corner back into view. (The solution is a table lamp with translucent shade compatible with the other lamp shades.)
- An incremental, confined light at the favorite reading chair. (The solution is an opaque floor lamp, positioned beside the chair with the bottom of the shade below eye level and equipped with a three-way bulb to provide low to high levels of illumination for quiet to busy activities.)

If lamps are the only light source for a space, observe the following rules of thumb.

Rules of Thumb for Lamps

- Never have only one lamp on in a space, unless the space is very small.
- Never read or perform other activities in a space with only one lamp lighted. (Additional light sources reduce glare, contrast, and shadows, permitting the eyes to function without excessive fatigue. All spaces should be without glare, but not all spaces should be without contrast and shadows. Contrast and shadows are desirable as long as they are not severe and harsh. Without them, living spaces would be dull and drab like a gloomy day. Overall, lamps should provide light that is adequate for seeing, diffused and distributed, glare-free, and free of severe contrast and harsh shadows.)
- Never position a lamp so that a shadow falls on the activity. (Heads, shoulders, and hands should not cause a shadow. The lamp should be placed so that it casts its light directly on the surface of the activity.)
- Have all lampshades at the same height, if possible. (This gives space a sense of unity. Other similar elements also do this, such as all the same metal and all the shades in the same unlit color.)
- For reading, the closer the light bulb is to the bottom of the shade, the more light will reach the reading material.

You need more light than you probably realize. The wide variety of lighting fixtures described in this book can provide that light, the easiest of which is lamps. However, use lamps with a lighting plan in mind.

An illuminated coffee table emphasizes a seating group.

lighting
a special
wall

5

Have you put money and effort into finishing one wall as a special wall? If so, you need to light it. Light permits the colors to be seen, emphasizes the design, and enhances textures. The light can either wash or graze the wall. Washing reveals the details and color of the carefully chosen wall covering, such as wallpaper, a mural, a wall graphic, wood paneling, fabric covering, or a contrasting paint color. Grazing reveals the intended texture, such as brick, stone, textured wallpaper, drapes, or other intentionally rough surfaces. Either type gives you the maximum benefit from the special wall, thereby justifying the expense. Further, light reflected from the wall adds to the other sources. Alone, it is sufficient for many occasions. Equally important, the reflected light flatters people in the space, and according to illumination research, people feel more positive about a space when the walls are lighted. Try it.

See for Yourself: Is Reflected Light Pleasing?

1. At night, in a room with a pale-colored wall, bring in one or two lamps with metal shades that throw the light in only one direction (for example, a reflector clamp-lamp or an auto trouble light).
2. Turn out all other lights.
3. Put the lamps close to the wall and direct the light toward it.
4. Observe the quality of the reflected light on people's faces and how the room looks.

Wall-washing is created by positioning the fixtures 1 ft (0.3 m) or more away from the wall and aiming them so that they cover it thoroughly and evenly. An owner can wash with a low-wattage, low-energy consuming system—either a system with low-wattage incandescent wall-washers or one with fluorescent cornice lights. A renter can use a portable wall-lighting system—either a system using an electric track with wall-washing track fixtures or one using portable wall-washers.

Wall-grazing, on the other hand, is created by positioning fixtures 6 to 8 in. (15 to 20 cm) out from the wall. The light creates shadows under any surface change, bumps, or depressions. The surface changes should be intentional, not accidental. An owner can graze a wall with recessed downlights. A renter can use a portable track system behind a cornice board. Both owners and renters can graze a wall of draperies with valance lighting.

Since light calls attention to what it illuminates and can make things visible that seemed invisible before, be careful what is lighted.

When Not to Light One Wall

- When the surface of the wall is imperfect and has unintended flaws, like bumps or nicks. Light exaggerates these flaws. On the other hand, intended texture, like textured wallpaper or fabric, is not disadvantaged.
- When the interior space is so narrow that lighting along one side could make it appear lopsided or create other odd visual or architectural effects.

- When you do not want to call attention to the wall, for whatever reason.
- When the wall covering is a mirror or has a glossy finish.
- When the wall is divided into small segments with small windows or doors.
- When the ceiling is sloped along the special wall, making the wall height uneven.
- When the wall is opposite large, undraped windows or glass doors.

When to Light One Wall
- When the wall is finished differently from the other walls.

- When you want to spread a small amount of light a long way.
- When you want to make a small room seem larger.
- When you have a very deep-colored wall, which recedes from view.
- When the wall is uninterrupted.
- When draperies cover the major portion of the wall.
- When you want to shorten an interior space, such as the end of a long hallway.
- When you want to balance the other lighting in the room.

wall-washing

Wall-Washing for Owners

Owners can choose either recessed incandescent wall-washers or built-in fluorescent cornice lights. Both can put the amount of light needed on the wall. Both function on regular 120-volt household current.

RECESSED INCANDESCENT WALL-WASHERS

For rooms with 8-ft (2.4-m) ceilings where a soft, low level of illumination similar to the glow of many candles is desired, use recessed incandescent wall-washers in the smallest size with 30- to 50-watt reflector bulbs. These fixtures do not attract attention to themselves. They fit into ceiling joist spaces that are 6 in. (15 cm) deep. The bulbs direct the light forward, and the fixtures redirect it onto the wall. They illuminate about 5 ft (1.5 m) down the wall (the visible wall above furniture). The closer to the wall they are positioned, the higher up on the wall they will light. Consequently, for graphics or murals that have details all the way to the ceiling, install the fixtures between 1 ft to 1 ft 6 in. (.30 to .46 m) away from the wall. For wall coverings that have details further down, install the fixtures 2 ft to 2 ft 6 in. (0.61 to 0.76 m) from the wall.

Likewise, fixtures need to be spaced symmetrically. Install them the same distance from each other on center—from the center of one fixture to the center of another—in an equal distant pattern, dividing the leftover space at each end equally. Thus, on a 16-ft (4.9-m) wall, seven fixtures could be spaced 2 ft 3 in. (69 cm) apart with 1 ft 3 in. (38 cm) at each end. Be sure measurements are made from the center of one fixture to the center of another.

For building, provide wiring for the wall-

washers when the electricity is roughed in. For relighting, recessing fixtures is possible if ceiling space permits, if wiring is accessible, and if fixtures can be secured to the ceiling material. In many of my relighting jobs, fixtures from one company have been adapted to accept the retaining clips from another in order to secure the fixture to the ceiling. Some companies are more

Wash a wall with little recessed wall-washers.

aware of relighting needs than others. Nonetheless, adaptations can be made.

The lighting effect is a series of gentle scallops are created at the top of the wall where the light from one fixture arches to meet the light from another. A dimmer is not necessary; the system allows three choices of bulbs for a little more or a little less light. When you want low-keyed illumination for quiet conversation, viewing television, or feet-up relaxing, use this system as the only light source. At other times when brighter light is desired, lamps and other sources should be available in the room.

For soft-light wall washing on pale-colored walls, use low-wattage BR (reflector) or compact fluorescent sources in well-engineered wall-washing fixtures. BRs produce from 380 to 1100 lumens.

Sample Electric Cost

On a 16-ft (4.9 m) wall with seven fixtures of 23-watt BR sources, the wall-washing system requires 161 watts. With lights on every night for 4 hours at 10¢ per kilowatt-hour, it costs $1.93 per month. The light will be soft.

If A-source fixtures are chosen, reduce the need to replace sources frequently by using a halogen-filled source (BT, TB, or MB) in frosted, not clear glass. It replaces any A source with three times longer life and is 60 or 100 watts.

When wanting greater brilliance or for rooms with either high ceilings or dark walls, use halogen QT-4 for the smoothest wash, or Krypton-filled or energy-saving reflectors (R) for soft-edge light. Or use halogen PAR or MR floods for a harder edge. PAR with an infrared reflective film is cooler and more efficient. Wattage for PAR ranges from 45 to 120, and from 20 to 75 for the MR. Halogen PAR is line voltage; MR is low voltage, requiring a transformer. Beam spreads range from 40 to 60 degrees. PAR-30 has either a short or long neck; and PAR-38 is large with the widest beam spread. Assure that the wall-wash fixture is highly engineered and is wide and deep enough to accept the source chosen without creating glare or wasting light trapped inside the fixture.

BUILT-IN FLUORESCENT CORNICE LIGHT

Fluorescent cornices are excellent for all styles of interiors and consume less energy. They are excellent for period-style interiors and traditional furnishings, which are normally difficult to combine with fluorescent lighting. They are

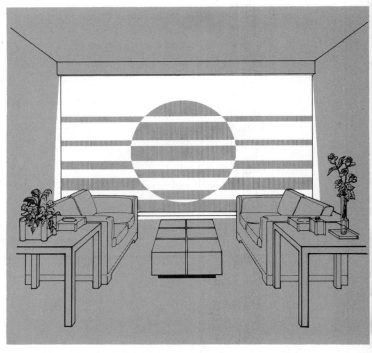

Wash a wall with a built-in fluorescent cornice light.

also sleek and tasteful in contemporary interiors.

A cornice light, if properly installed, confines all of the light and does not attract attention to itself. It focuses on the wall. This system can accommodate ceilings that slope away from a wall, provided the wall height is even. The fluorescent light goes about halfway down the wall, trailing off in brightness.

If incandescent sources are in the room, choose warm color of light (30 or 35K fluorescent); otherwise, the colors of light will fight. Use all the same Kelvin temperature sources in the cornice to assure uniform color of light on the wall. Also, the color of light should flatter the occupants. Typically, in our culture, warm colors of light appeal. (Far Eastern cultures think differently.) Preferably, use T-8 rather than T-12 sources. They are thinner and more energy efficient, and have higher color rendering. But T-8 requires an electronic ballast to do so.

Linear fluorescent sources offer a wide variety of choices. Size choices are T-12 at 1½ in. (38 mm) diameter, T-8 at 1 in. (25 mm), T-5 at ⅝ in. (16 mm), plus TT-5, twin-tube, long compact sources at ⅝ in. (16 mm). TT-5's are excellent for small, recessed, ceiling-mounted fixtures giving a smooth wash.

Be aware that only the 3-ft and 4-ft (0.9-m and 1.2-m) rapid-start fixtures can be dimmed. The instant or preheated fixtures cannot. However, dimming is usually not necessary, because fluorescent light is already low in wattage. Likewise, fluorescent fixtures do not usually create unpleasant brightness if the installation is designed well.

All fluorescent fixtures are distinguished by the quality of their ballast, but the quality is hard to determine. Rely on the rating of several organizations that test ballasts. The best ratings are a Class P for safety, a CBM certification for quality standards, and an A rating (on a scale of A to D) for the quietness.

A cornice system can illuminate evenly only a wall that is uniform in height. Therefore, the ceiling must be either flat or evenly sloped away from the wall. The farther the fixtures are installed from the wall, the farther down the wall the light will be thrown. In fact, for each 1 ft (0.3 m) out from the wall, the light will go down 4 ft (1.2 m). If you want light on the visible wall above furniture, install the fixtures 1 ft (0.3 m) out.

Place the fixtures end to end, within the limits of the fixture sizes. Normally, some empty space is left over, but keep it to a minimum and equalize it at each end. Or, if more than a minimum space is left, equalize it by separating each fixture by an inch or so, using up the extra inches. However, be aware that empty spaces between fixtures create shadows that destroy the smooth-wash effect.

Use this installation method:

1. Make a cornice board of hardboard or plywood faceboard, no less than 6 in. (15 cm) deep and as long as the wall being lighted.

2. If necessary to prevent the tubes from being seen from a usual position in the room, seated or standing, a narrow return or piece of wood can be added to the bottom of the board. In addition, the return could hold a diffuser or louver to further hide the tube. Neither is necessary in most installations.

3. Paint the inside of the board flat—not glossy—white.

4. Attach the board to the ceiling at least 6 in. (15 cm) out from the wall.

5. Mount the fixtures on the ceiling next to the board either on 1- by 3-in. (2.5- by 7.6-cm) wooden blocks or directly behind the board if decorative molding is to be attached on the outside at the top. Either way, the light cannot leak out at the top of the cornice board.

6. The outside of the cornice board can be painted, stained, covered with fabric or wallpaper, or plastered. Use whatever enhances the style of the room and blends with the wall covering.

7. Modify the measurements of the depth of the cornice board by the proportions of the board to the space, the view under the board, and other architectural considerations. Also, modify the measurements of distance from the wall by considering how far you want the light to be thrown down on the wall.

When building, a cornice system can be preplanned and the wiring provided. The cornice board can be made to match the other woodwork in the space, especially if it is distinctive.

When relighting, a cornice system can be installed if wires are available either from access through the attic, from a hot switch, or from a baseboard receptacle. Skilled electricians often can reach wires in wooden frame construction, but not so often in other types of construction.

The light will spread as far as the fixtures are spread—ideally from end to end. More light will be at the top of the wall than at the bottom, but it will be bright and pleasing. The amount reflected into the room will be rich and glowing, sufficient to read the weather forecast, but not enough to read the whole newspaper.

Sample Electric Cost

A fluorescent system for a 16-ft (4.9-m) wall using four 4-ft (1.2-m) tubes of 40 watts each requires 160 watts for the tubes and 24 watts for the ballast. Used for 4 hours per night at 10¢ per kilowatt-hour, it costs $2.21 for a month. Fluorescent bulbs last up to 20 times as long as incandescent bulbs, and they spread light farther. In addition, they create a lot less heat. The overall saving—fewer watts for lighting and less cooling by fan or air conditioner in hot weather—would offset the slightly higher installation cost of the fluorescent system.

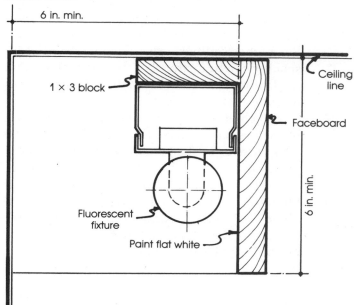

Cornice details.

Wall Washing for Renters

Renters can install a temporary wall-washing system consisting of a plug-in track and line-voltage or low-voltage track fixtures. The track and fixtures become a decorative element in the space. They attract attention to themselves, but also put attention-getting light on the wall.

TRACK AND TRACK FIXTURES

Track fixtures accept a range of sources and wattages for wall washing. Line-voltage incandescent BR (reflector) spots and floods in R-20 or -30 sizes offer 30 to 100 watts. Low-voltage MR-16 offers narrow spots to wide floods from 20 to 75 watts (requiring a transformer). Compact fluorescent offers 13 to 40 watts using compact or linear sources. They equal the light of 100- to 150-watt incandescent, with 70 percent less electricity. Metal-halide T-6 offers 70 watts. What an array of choices. Choose the type based on how much light is needed from the sources for the least energy.

Track fixtures throw light forward. They are adjustable and convenient for installation uncertainties—changes of mind, timid installation calculations, or refining the aiming angle because a ceiling slopes away from the wall.

Install track sections parallel to and as long as the wall. Do not use a short track and try and "throw" the light obliquely to the wall. It will be uneven and could be glaring. Use a track long enough to have fixtures evenly spaced. The distance to the ends of the track should be half the distance between the fixtures. Typically, for 8 ft (2.4 m) ceiling heights, fixtures can be spaced between 2 and 3 ft (61 and 91 cm) apart and the same distance to the wall. For higher ceilings, track distance and fixture spacing can be farther, but the farther back and the farther apart, the less light. Higher ceilings definitely require greater lumens.

If chairs, sofas, and other seats are in front of the wall, make sure the light will not shine into the eyes of the person seated. If in doubt, position the track closer to the wall.

The wall will be washed smoothly with moderately intense light, which will be sufficient to illuminate dark wall coverings, like wood paneling and deep tones of textured grass cloth. On paler walls, the amount of light re-

Wash a wall with track lights.

Renters can install a temporary wall-washing system of uplights.

flected should be cheerful and sufficient to enable one to enjoy activities with simple tasks, such as putting photos in the family album.

PORTABLE WALL-WASHERS

Portable wall-washers permit renters to take advantage of the benefits of wall-washing without disturbing the structure or the terms of the lease. They stand on the floor and connect with a cord and plug. However, they light from below. For these reasons, they are highly visible and make a definite impact. They are decorous for any interior that would not be disadvantaged by many floor fixtures—usually a contemporary or eclectic one.

Portable wall-washers are either canister-shaped, about 12 in. (30 cm) high and 6 in. (15 cm) wide, or a track fixture mounted on a weighted canopy base. Each must be equipped with a switch and must be turned on independently, unless the wall receptacles are wired in series (together) and can be controlled from one wall switch. A switch allows a dimmer to be installed. Use a reflector bulb in a 150-watt size, a 90-watt PAR, or a standard A bulb in 90- to 150-watt size.

All portables must be positioned evenly. Typically, the distance is about 3 ft (0.9 m) from the wall and about 3 or 4 ft (0.9 or 1.2 m) apart for the whole length of the wall, making the end

Portable uplight.

spaces about half the distance between the fixtures. Experiment with your wall to discover the best positions.

With these portables, the light originates from the floor. Therefore, the brightest part of the wall is at the bottom. Wall coverings containing details down to the floor can be well illuminated. Wall coverings with details farther up should have the fixtures positioned farther back. Floor fixtures can be hidden behind furniture in front of the wall.

Sample Electric Cost
On a 16-ft (4.9-m) wall, five portable wall-washers with five 100-watt bulbs, used 4 hours per night at a utility charge of 10¢ per kilowatt-hour, cost $6 per month.

---

wall-grazing

Wall-grazing is created by aiming light at a shallow angle to the wall. The fixtures must be positioned very close to the wall. The light enhances any texture, or other surface qualities, all of which should be intended. If the wall has unintended flaws, the light calls attention to them and they become predominant. Do not graze plaster or drywall that is only painted. Graze a wall covered with textured wallpaper, fabric, or masonry—particularly where the mortar has been removed from the edges (raked joints). A masonry wall stands up and sings with grazing light.

Certain requirements must be met to produce grazing light:

• The fixtures must be close to the wall.
• Incandescent reflector bulbs must be used. Standard A bulbs do not concentrate the light enough, and fluorescent is too shadow-free.

• The light bulbs must be well concealed, especially from view when seated.
• The fixtures must be close together.
• The fixtures must be out of sight and unimposing, either as recessed fixtures or track fixtures behind a cornice board.
• To produce the greatest amount of light with the least electricity, a recessed fixture must have a built-in reflector, and a cornice board must be painted flat white inside. Recessed fixtures without internal reflectors will graze a wall for a shorter distance, such as above the mantle over a fireplace.

When the fixture is positioned close to the wall, a pattern of scallops is created at the top of the wall where the scallop of one light meets the scallop of another. For some people, the scallops heighten the dramatic effect. Others disagree. If it is desired, the scallops can be virtually eliminated by positioning the fixtures very close to each other. However, the energy consumption and the heat are increased. In some installations, these disadvantages are outweighed by the look of the wall.

See for Yourself: Do You Object to Scallops?
1. Take two flashlights to a plain, pale-colored wall.
2. Light the flashlights and hold them about 12 in. (30 cm) apart, with your arm against the wall, one near your elbow and one in your hand.
3. Observe the scallops.
4. Note that the farther apart you hold the flashlights, the deeper the scallops become.

Wall-Grazing for Owners

Owners who are building and relighting can use recessed, multiplier-type downlights attached to ceiling junction boxes. The multiplier-type fixtures have internal reflectors that focus all the light and redirect it down. The reflector, sometimes called a cone reflector, is made of highly polished metal. Some reflectors are not well designed and show an image of the light bulb on the reflector. Choose one that does not. Inspect reflectors in your local showroom.

The greater the wattage, the more intense the shadows. The wall color affects shadows. Dark walls obscure shadows. On pale walls with intense light, shadows can be seen. A 50-watt, MR-16 narrow flood in a well-engineered fixture could graze a 12-ft (3.7 m) high wall. So could a 75-watt, halogen, PAR-30 flood, preferably infrared.

Install the fixtures 6 to 8 in. (15 to 20 cm) out from the wall and about 16 in. (41 cm) apart, evenly spaced. The further the spread, the deeper the scallops. The distance at the ends should be half the distance between the fixtures. Uneven spacing will be very apparent in wall-grazing. It will create uneven scallops and uneven light.

The system can be dimmed. Often, dimmers have been used where the initial wattage was too much for the space and the brightness had to be cut down to what it should have been in the first place. This practice is poor lighting design. Instead, dim to extend the life of the bulbs, especially if the installation is not in a convenient location to change bulbs. In addition, dimming changes the color temperature of light and emphasizes the red end of the spectrum. Therefore, dimmed light could be used to enhance, for example, red tones on a brick wall.

When you are building, plan for sufficient ceiling depth for wall-grazing fixtures. When you are relighting, the depth must be there already, and wiring must be gained through an attic or by pulling wires from elsewhere. If ceiling wiring is not obtainable, owners will have to use the temporary wall-grazing methods suggested for renters.

Wall grazing makes the wall texture very visible, enriching and enlivening the special wall. The light will be sufficient for conversing, listening to music, or viewing television.

Sample Electric Cost
Light for an 8-ft (2.4-m) wall using six fixtures of 50 watts each for 4 hours per night at 10¢ per kilowatt-hour costs $3.60 per month. The amount is no more than would be required for two table lamps, and yet two table lamps could not light a whole wall.

Wall-Grazing for Renters

Renters do not have the options that owners have to recess fixtures. Nonetheless, they can have dramatic grazing light by two means—either track fixtures hooked up with a cord and plug, or bare bulbs on a track behind a cornice board. Track fixtures have their own shielding, and can be installed to point almost straight down along the wall. Bare bulbs on a track need a cornice board to shield them from view.

To make a cornice board, follow the directions for the built-in fluorescent cornice in this chapter. However, mount the board between 8 to 12 in. (20 to 30 cm) away from the wall, depending upon the size of the track fixture. Rings or bare sockets are the smallest. PAR-38 rings are 5 in. (13 cm) wide and MR-16 rings are 3 in. (8 cm) wide. Check the angle of view from both seated and standing positions. The cornice board should be deep enough to obscure the sources. It should not be obvious where the light is coming from—mystery is part of the delight.

Make the board 12½ in. (32 cm) deep with a 3-in. (8-cm) return or less at the bottom. The track should be mounted on the inside of the board. The outside of the board can be trimmed, painted, or otherwise finished to fit and coordinate with your wall. Install a cornice board as long as the wall to be grazed, and adjust the track fixture to point down.

Use the simplest track fixture—a socket for the bulb—and equip it with a spot bulb. These fixtures are functional-looking and must be hidden. Each bulb should be 10 to 12 in. (25 to 30 cm) from the other and 6 in. (15 cm) in from the end. However, different spacing or type of bulb

Graze a textured wall with recessed lights.

might be more satisfactory for your wall. See for yourself by checking the effects.

Choose the wattage of the source in relationship to wall color and ceiling height. The wall color will reduce the light by the reflectance percentage of that color. (Dark is a greater percentage; pale is less.) Also, the ceiling height will reduce the amount of light for casting shadows by the square of the distance away from the source. It is not a linear reduction, but a logarithmic reduction—hence greater! Thus, for dark walls and high ceilings, choose the highest wattage MR (75) or infrared PAR (100) sources possible and pay the electric bill, but feast your eyes.

The precise alignment and wattages need not be determined before installation. Changes of spacing and wattage can be made afterward. Often changes are required in both commercial and residential spaces.

The wall will be brighter at the top and softly lighted down the wall, depending upon the wattage chosen. For example, the effect of using the 75-watt spot is a strong stroke of light touching the wall all the way down. It puts out enough light to let you glance at a magazine or find your shoes easily, but not enough to read a fifth carbon copy of an order.

Sample Electric Cost
A cornice system using eight 50-watt fixtures along an 8-ft (2.4-m) wall for 4 hours at 10¢ per kilowatt-hour costs $4.80 per month. On the other hand, a custom-made system equipped with 75-watt spots, spaced 12 in. (30 cm) apart costs $7.20 per month. Even with the higher wattages, the yearly additional cost of electricity for a wall-grazing system is about the same as four tanks of gas for the family car or one for the office delivery truck.

grazing window walls and draperies with a valance

Often, large windows take up a whole wall or part of a wall. Sometimes these windows are covered with draperies, which are usually closed at night. The draperies become a major element in the space, contributing color and texture. Lighting the draperies not only reflects light from the wall but also bounces it off the ceiling, giving the space greater illumination. Light the drapes with a valance. Valances can be used easily with sloped ceilings because they are installed well below the ceiling line and are not affected by an uneven wall.

A valance is basically like a cornice but is installed 10 in. (25 cm) or more down from the ceiling. The faceboard can be made the same depth as a cornice board, but it is attached to the wall. (If the board were installed at the ceiling, it would be called a cornice board.)

Certain requirements must be followed to install valance lighting:

- The faceboard must be installed a minimum of 10 in. (25 cm) from the ceiling to let enough light out. (A greater distance is even better light on the ceiling.)

- The light source must be at least 2 in. (5 cm) away from the faceboard and the draperies to dissipate the heat.

Valance lighting on draperies bounces light into the space.

- The depth of the faceboard must be at least 6 in. (15 cm), but usually needs to be deeper to conceal the light source from people seated.

- The board must be painted flat white inside to reflect as much light as possible.

- The board must be at least 4 ½ in. (11 cm) out from the wall for clearance.

- The light source must be at least 2 in. (5 cm) from the bottom of the faceboard to be hidden.

- The light should be smoothly spread for the whole length of the draperies. Any distance not lighted at the ends should be equalized.

- The light source can be a fluorescent or remote-source lighting with an MR-16 source in an illuminator box and prismatic film to spread the light.

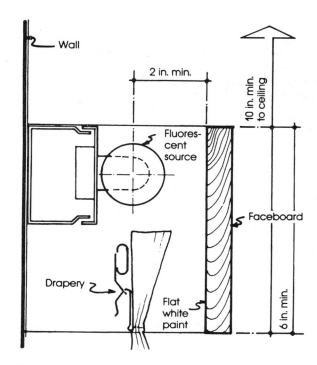

Valance details.

The light from a valance can be soft and appealing. It spreads as wide as the fixtures are spread and appears above and below the valance. It permits short-term, easy-to-see activities in residences. Or in commercial spaces, it gives a broad area of light without dark corners.

Valance light can be fluorescent or incandescent, either with a remote-source fixture or a track system. A fluorescent valance system consists of fluorescent light fixtures behind a valance board. Install the same way as a built-in fluorescent cornice described in this chapter. Choose between linear fluorescent T-12, T-8 with electronic ballasts, T-5, or TT-5 (long compacts). They vary in diameter and in amount of light produced (lumens). T-5 is the slimmest.

An incandescent, remote-source system is composed of an MR source in an illuminator box, remote from the drapery in a well-ventilated enclosure, a light guide of hollow prismatic film, or into a light guide of end-lit fiber optics with curtain-washer fixtures. Such light is energy efficient, cool, and uniform. (See Indirect Remote-Source Lighting in Chapter 9.)

Owners who are building or relighting should use fluorescent or remote-source lighting. Renters will want to use incandescent track fixtures temporarily hooked up with a cord and plug behind a ready- or custom-made faceboard. Ready-made boards might not be large enough to shield the needed track heads. Custom-made boards could be.

If the draperies do not cover the whole wall end to end, the faceboard must have end pieces to obscure the light fixture and complete the valance. The faceboard can be held to the wall with ½-in. (13 mm) metal angle brackets or wood brackets. The board can be finished to harmonize or contrast with the other interior finishes of the space.

Some manufacturers make ready-made, fluorescent brackets that can be mounted just over the window or go from wall to wall. They can be customized with special colors. Most have lenses below to obscure the light sources.

For those who do insist on rejecting fluorescent sources and cannot afford a remote-source system, linear incandescent sources can be used, but they require more electricity and light source replacement, and generate more heat. With linear incandescent, like fluorescent, their length is tied to their wattage—12 in. (30 cm) at 35 watts, 20 in. (51 cm) at 60 watts, and 40 in. (102 cm) at 120 watts. They can be mounted end to end behind the faceboard and can be inexpensively dimmed, but cost more in the long run.

Sample Electric Cost

For an 8-ft (2.4 m) window, a linear incandescent system would consume 240 watts and a linear fluorescent system would consume 64. At 4 hours per night, the incandescent system would cost $2.04 more per month than the fluorescent at 10¢ per kilowatt-hour.

The amount of illumination seen ranges from bright to soft, depending upon the sources used, the color and texture of the draperies, and the ambient light in the rest of the space. Research has shown that people perceive brightness in a space if the walls are bright. Use this application to increase the perception of brightness and enhance the fabric chosen for the draperies.

downlights
for settings
6

Downlighting is one of the best ways to light. Downlights can create many settings for homes or businesses that are not obtainable by any other type of lighting. Downlights can create sparkle, and can carve out, expand, and punctuate a space. They can provide a lot of or a little light that harmonizes or contrasts. With downlights the ceiling is not bright; surfaces—tabletops, walls, furniture, and floors—are bright. Consequently, surfaces attract your attention, not an uninteresting ceiling.

See for Yourself: Do Surfaces Attract Attention?

1. In a dark room, turn on one lamp with a translucent shade.
2. Observe that your attention is drawn to the lampshade.
3. Open the bottom and top of a box big enough to cover the whole lamp, yet allow the light to go to the ceiling.
4. Observe that your attention is drawn to the ceiling surface and you are unaware of the lamp at all.

Some downlights are well-engineered fixtures. Choose them to get the most light for your money. Some downlights put the light straight down; some put it to the side. How the fixture looks does not always indicate the kind of light it produces. Do not try to guess what it does.

In theaters, settings are created by skillful placement of well-engineered lighting equipment. In commercial spaces, settings are created by downlights to sell products. Likewise, settings can be created in residences to emphasize, soothe, or brighten. What kind of setting do you want?

sprinkle a horizontal surface with sparkle

Create a brilliantly sparkling tabletop with light. It is attention getting and cheerful. The colors and objects on the table become vivid and appealing. Contrast the setting to a more subdued background. Make the downlight bright.

The setting draws people to it. The closer the light is confined to the tabletop, the more intimate it becomes. This effect is good at home as well as at the store.

Two ceiling downlights or a downlight in a

chandelier can provide the kind of setting. Source choices are halogen PAR, MR-16, or T-4. PAR and MR provide the most sparkle for the energy consumed. Those that have the infrared technology produce more light (candlepower). Thus, an infrared 50-watt MR-16 spot source could provide 800 more candlepower than a non-infrared MR-16 for the same energy consumed and the same life span.

COMPARISON OF MR-16 PERFORMANCE

TYPE	BEAM	CANDLEPOWER	WATTS
infrared	10°	16,000	50
non-infrared	10°	15,200	50

Deliver more candlepower (19,000) by using a low-voltage source in a line-voltage screw-base with an internal transformer (called an Exhibit PAR-38).

carve out space within a space

A space can be defined by downlights. The light need not be brilliant, but the surrounding light must be dimmer in order not to loose the integrity of the carved-out space. Brighter illumination could be further away if it did not interfere.

For example, downlight, focused on comfortable furniture inviting you to sink-in, visually organizes the seating area with consistent light. The light creates seemingly three-dimensional but penetrable walls. It is soothing and yet sociable. The illuminated area seems to be alone and suggests you sit back, put your feet up, and forget the cares of the day.

In residences for a softly illuminated area, put up to 23 footcandles on the floor from an 8-ft (2.4 m) ceiling by using 32- or 42-watt compact fluorescent sources. For a vividly illuminated area, use 250 footcandles by using halogen T-4 sources at 150 watts. In between, use halogen PAR-38. Position a downlight over the front or back edge of a seat, never over people's heads. Overhead produces shadows, not flattering to anyone.

A softly illuminated seating area of one

Carve out a space within a space with downlights.

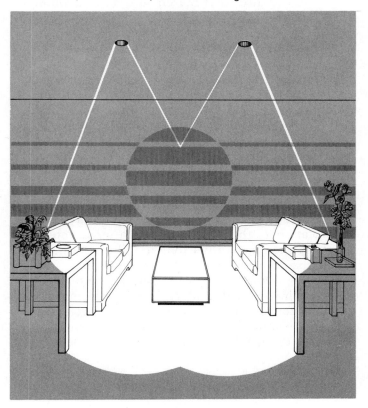

small and two large sofas, and a coffee table would require three 30-watt downlights. Position the fixtures over the front edge of the cushions at the ends of the large sofas and over the middle of the small sofa. In addition, position a 50-watt downlight over the coffee table to fill the interior of the area.

Sample Electric Cost
The electrical cost for one 50-watt and three 30-watt downlights for 4 hours per night for a month at 10¢ per kilowatt-hour would be $1.68.

expand a space
with widespread light

A room can be expanded by light, exploding the space and fostering an atmosphere that encourages people to move around freely. Light reveals the richness of the whole space. It is stimulating. Several fixtures are required to spread the light uniformly. The amount of light from each fixture should be the same, and the fixtures should be spaced evenly. When choosing a fixture, consider personal preference about the fixture's aesthetics, and coordinate its light-delivery capabilities. Wide-beam downlights can do the job. Fluorescent fixtures can also work, but they create distractingly bright patterns of light on the ceiling. Unfortunately, they are overused in commercial spaces. Parabolic wedge louvers on the fixtures will make them less obvious. However, substitute downlights in some spaces and provide visual relief from the rows of fluorescent fixtures that march across most commercial ceilings.

When using incandescent downlights to expand a space, create uniform light by positioning the fixtures evenly throughout the space. The distance between fixtures can be determined easily with spacing ratios. However, if a specific footcandle level must be delivered, more complicated calculations will have to be made. (See calculations in Chapter 21.) To easily determine spacing, either follow the manufacturers' spacing ratio, or position the fixtures apart by the same distance they are from the floor. For example, an 8-ft (2.4-m) ceiling should have fixtures 8 ft (2.4 m) apart. If fixtures are too far apart, the light becomes scattered, chopping up the space rather than unifying it. If they are spaced correctly, the light is uniform.

Manufacturers indicate spacing ratios in their technical catalogs. They are the maximum distances their fixtures can be apart and still deliver uniform light. A spacing ratio multiplied by the height above the surface to be lighted (usually the floor, possibly a tabletop) equals the space between fixtures. The space between a fixture and the wall should be half the distance between the fixtures.

Rule of Thumb for Spacing Fixtures
Spacing = Spacing ratio × height above the surface
to be lighted

Spacing ratios range from 0.3 to 1.4. The greater the ratio, the farther apart the fixtures can be installed. For example, in a room with 8-ft (2.4-m) ceilings, fixtures with a spacing ratio of 1 should be centered 8 ft (2.4 m) apart; fixtures with a ratio of 1.4 should be centered 11 ft 2 in.

Expand a space with a widespread light.

(3.4 m) apart. Of course, the actual installation distance must be modified by the size of the room, because room sizes cannot always be divided evenly by spacing ratios. As you modify the spacing, be aware that the closer a fixture is to a wall, the more likely a scallop of light will be formed on that wall. Sometimes the scallop can illuminate something you do not want seen, disturbing an interior setting. Determine any disadvantages before installation.

punctuate a distinctive small area
with a pool of satiny light

A small bay window in a residence or a low-rise commercial structure can be illuminated. From the outside, the lighted bay window greets and says goodbye. At night such a light can be a nightlight, giving a reassuring glimmer. When the room is lighted and full of people, the downlight balances the other illumination. The downlight need not be bright. If the surrounding light is not overpowering, 30 watts can be bright enough. If the surrounding light is bright, 50 or 75 watts may be necessary.

Sample Electric Cost
Only 36¢ worth of electricity is used for a 30-watt downlight operating 4 hours per night for a whole month at 10¢ per kilowatt-hour.

Many other settings can be formed with downlights. Determine what is to be emphasized and how vivid it should be. Of course, the more vivid it is, the more attention it will draw. But not all settings require brightness, nor should they be bright. Remember, vivid light is stimulating; soft light is soothing. Decide which mood you want.

In addition, the mood can be changed with a dimmer—a device to reduce the light output of the bulb. As the incandescent light dims it also turns more golden in color and gives the appearance of being warmer, enhancing the red colors in the space, but sometimes making other colors appear muddy.

From the outside, a lighted bay window greets and says goodbye.

Punctuate a distinctive small area with a pool of satiny light.

effects

The lighting effects for settings range from broad to focused, from harmonizing to contrasting, and from soft to vivid. When the whole room is illuminated, a social, gregarious atmosphere is created. When an intimate space is illuminated and the rest of the space is dark, people are drawn to the lighted space. When the walls are illuminated, a self-contained, soothing atmosphere is created. The effect of downlights depends not only on the engineering of the fixture, but also on the type of incandescent bulb used. Some downlights are engineered with reflectors to concentrate the light; others spread the light. Fixtures that put light straight down produce wide, medium, or narrow lighted areas. The appearance of the fixture does not always indicate the width of the light it produces. For example, a pinhole looks like it produces a narrow beam of light, but it does not. It produces a medium beam if the bulb is correct and properly adjusted. A narrow beam is produced by an open fixture equipped with a very narrow spot bulb. Do not try to guess the amount of light a downlight produces; consult the manufacturer's catalog, a knowledgeable salesperson, or a lighting consultant.

fixtures

The engineered features of the fixture are internal or external. Internal features are reflectors and adjustments. Reflectors redirect, spread, or concentrate the light from within the fixture. Some reflectors are better designed than others. The best reflectors enhance the light and do not reflect the source's image at usual viewing angles (45 degrees). In general, cone reflectors redirect the light, especially from incandescent-A sources. Cones are specular, clear (silver), gold, or black. The color does affect the amount and color of light.

Usually, elliptical reflectors minimize the light source and focus the light through a small aperture for a medium to broad distribution.

Similarly, parabolic reflectors redirect the light, but through a wider aperture. They are best when combined with black baffles to catch the stray light at the aperture.

The second internal engineering feature is adjustment. Adjustments permit the light source to be reaimed for accenting or compensating for a sloped ceiling.

External features are lenses, diffusers, baffles, and louvers. Lenses cover the opening of the fixture and bend the light in some way. A Fresnel lens concentrates the light. A spread lens redirects it to the left and right. A prismatic lens directs the light down, but it can also be bright. To make the best use of a lens, the bulb needs to be the exact wattage specified by the manufacturer. Then the light will be precisely focused. The engineering of a downlight can be ruined by the wrong light source.

Diffusers obscure the source and scatter the light. Opal diffusers are the best to use. When they are the only sources of light, they are bright, attract attention, and in some situations are distracting. Smooth out the brightness with additional light sources in the space.

Baffles (black grooves in the inner surface of the fixture's opening) and louvers (metalwork over the aperture) also reduce aperture brightness. Some louvers are unidirectional, intended to throw the light in the direction they are pointed. However, these are not as effective as internal reflectors.

All downlights can be described in terms of how they distribute the light: wide, medium, or narrow beam; adjustable accent; wall-washer; framing projector; or downlight/wall-washer.

A wide-beam downlight is generally used to illuminate a large area. A medium-beam downlight is used to illuminate an area where

Reprinted by permission of Lightolier.

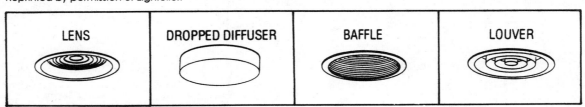

the activity and the objects to be viewed are important—a tabletop, a seating area, or around other furniture. (On the other hand, the space becomes more important and the activity becomes less important when walls are illuminated.) A narrow-beam downlight is used to highlight special areas or objects. In rooms with high ceilings, many narrow-beam downlights can illuminate a whole room. Also, both wall-washers and adjustable accent fixtures are engineered to put light to one side. (Wall-washers are discussed in Chapter 5.) Adjustable accent fixtures can point to the side, or they can shoot light straight down from a sloping ceiling, com-

Wide.

Medium.

Narrow.

Wall-washing.

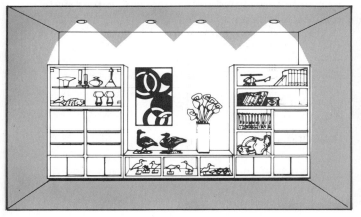

Adjustable accenting.

Framing projector.

Downlight/wall-washer.

pensating for the slope. Some are internally adjustable; others are externally adjustable. A framing projector confines the light to a precise, sharp-edged beam. (Consult Chapter 8.) Finally, a downlight/wall washer lights two different surfaces at one time. It permits some light to fall on the floor, as well as some directed to the walls, or even a corner.

Some downlights are not well engineered. They are essentially tin cans with sockets. Well-engineered fixtures are not necessarily the most expensive in the long run. Well-engineered fixtures are efficient and provide more light for the watts consumed. They are rated by an efficiency percentage, which is a measure of how many lumens the fixture puts out compared to how many lumens the light source generates. The higher the percentage, the more efficient the fixture.

Some downlights can be covered with insulation—IC label. Some downlights cannot be covered—TC labeled—and they have a thermal cutoff to avoid overheating. Some downlights are airtight and reduce toxic radon gas from entering structures when air escapes through ceiling downlights. Further, airtight fixtures reduce room-to-room noise transfer and intrusion of outside noises (like airplanes overhead).

Fixtures are manufactured to effectively utilize the source. Follow manufacturers' specifications for source type and wattage. Do not try to outwit the engineering of a fixture.

Light source choice dictates apertures of downlights. Smaller sources mean smaller downlight apertures, except for the pinhole, which uses R, PAR, or MR sources.

light sources

Downlights accommodate a wide range of sources—incandescent, compact fluorescent, or high-intensity discharge (HID)—both metal halide and sodium. The watts range from 9 to 500. Many line-voltage downlights accept more than one wattage and type of source. These fixtures internally adjust to ensure that the source is correctly positioned to deliver the most light. Otherwise, light gets wasted. Further, many are deep enough to accept a reflector, screw-base, compact fluorescent to reduce energy consumption. Never replace without a fluorescent reflector. It redirects the light from the linear source.

Narrow beams come from incandescent PAR, reflector (BR), or MR spot sources and from fixtures designed to deliver narrowly. Wide beams come from incandescent A, PAR and BR floods, and HID sources.

When choosing between the sources, there are trade-offs. A's are less expensive to buy, but burn out quicker. The widest coverage from A is with an ellipsoidal reflector inside the fixture.

PAR is more expensive to buy, but last longer. Halogen PAR produces more light than BR and other PAR's. T-4 needs more wattage to produce the light than a PAR does.

Compact fluorescent requires less wattage than A sources for the same spread of light. In addition, a compact-fluorescent fixture can have an emergency ballast to provide light for escape routes, power outages, and other safety reasons.

Residences need them as well as commercial spaces. When lights go out at night, homes are affected.

Metal-halide needs more wattage to produce the amount of light white sodium does. Both need time to cool down before turning on again. Thus, they are more suited to commercial, rather than residential spaces.

COMPARISON OF PERFORMANCE

SOURCE	WATTS	fc @ 8 ft	BEAM WIDTH @ 8 ft
Incandescent:			
A	100	9	10½ ft
PAR-20/NFL	50	26	4 ft
T-4	100	28	5 ft
Fluorescent:			
Triple Compact	32	8	8 ft
HID:			
Metal-Halide	100	31	9 ft
Sodium	50	28	8½ ft

White sodium sources for downlights are an excellent choice for commercial spaces. They render colors well. The 35 watt produces as much light as a 100-watt incandescent. Likewise, it produces the least amount of ultraviolet light of any artificial source. And ultraviolet definitely fades surface colors.

In addition, incandescent elliptical reflectors (ER) have narrower beams than floods, but wider than spots. The overall length is long and requires very deep recessed fixtures.

Further, compact fluorescent sources are suitable for downlight, if the cone reflector of the fixture does not reflect an image of the source. Compact brand names are PL, Dulux, and Biax. By any name, they are energy efficient and long-lasting. Triple compact fluorescent sources are the most brilliant energy efficient and last the longest. Wattages for fluorescent downlight fixtures vary from 18 to 42, and configurations vary from two to four bent tubes. Some sources are manufactured with less mercury so that they can be disposed of without treating as hazardous waste. Compacts have a full range of color temperatures—30, 35, and 41K. Some with electronic ballasts are dimmable. And some can be used at freezing-level ambient temperatures.

Choices of sources abound. The difficulty is deciding which one will do the job the best!

All compacts are bent glass tubes, but differ in shape and number of tubes (two or four). Some are long and thin; others are short and fat. Some tubes (usually the 9- and 13-watt sizes) are in a globe or in a glass reflector with a built-in ballast. Bases vary. Bases, watts, and ballast requirements must be matched. Do not mix.

The 18- to 27-watt sizes are good for a 1 x 1 ft (.3 m) fixture. The 39-watt size fits a 2 x 2 ft (.6 m). The long and thin T-4 is suitable for higher ceilings when higher levels of light are required (usually 18 and 26 watts). But, it is not as good for retrofits. On the other hand, the short and fat T-5 is good for retrofits (usually 22 and 28 watts), requiring only a simple ballast. But the light output is less. The four-tube compacts put out the amount of light that 40 to 100 incandescent watts do. Overall, compact-fluorescent sources offer many options and create a warm light which blends with incandescent. Fluorescent downlights are excellent for producing low-ambient light and for not adding heat to strain air-conditioning.

The high-intensity discharge (HID) source for downlighting is the white sodium T-10 in 35 to 100 watts. This excellent source produces good color of light like incandescent, renders colors well, and produces more light per watt (lumens) than incandescent. The 35-watt version produces as much light as a 100-watt incandescent. Likewise, it produces the least amount of ultraviolet light of any artificial source. Ultraviolet fades surface colors in any space.

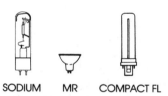

SODIUM   MR   COMPACT FL

Reprinted by permission of Philips Lighting Division of N.A. Philips and Sylvania.

installation

All downlights must be connected to the electricity, either through a ceiling junction box or by way of a track. Through a junction box, fixtures are installed recessed, surface-mounted, or semirecessed. The fixtures are hooked up to the electricity within the ceiling. Recessed fixtures require sufficient space to put the fixtures above the finished ceiling. The recessed fixtures are the least conspicuous, and they enhance the spontaneous effect of the setting. In wooden frame construction, the ceiling joists usually are 16 in. (41 cm) on center, leaving available 14 in. (36 cm) in width and 10 in. (25 cm) in depth. However, if the available space is not sufficient, semirecessed (dropped partially below the ceiling) or surface-mounted downlights might be possible.

The National Underwriters' Code requires that recessed downlights either be thermally rated—engineered to permit covering with ceiling insulation—or have a thermal switch in case they get too hot. Before thermal protection requirements, all downlights were installed 3 ft (0.9 m) away from insulation. During that time, one of my clients asked me to determine how much heat and air would be lost with ceiling downlights if the insulation had to be 3 ft (.9 m) away. After consultation with another university and the State Energy Office, I gave him a determination. We did not have a chance to verify the calculations because he bought a house and did not build. Now, thermal-rated downlights can be covered with insulation, and they will prevent loss of heat or air-conditioning.

In concrete and steel construction, tracks can be surface-mounted and plugged into a baseboard receptacle, with track fixtures installed as downlights. In all construction types, tracks also can be hooked up to ceiling junction boxes. In commercial spaces with suspended ceilings, downlights are easy to install.

Owners can install recessed fixtures in new construction if wires are provided. When relighting, owners must have an electric junction box in the ceiling, or the wires can be brought in through the attic. In addition, the electricity must be controlled by a wall switch. A dimmer can be installed to give greater flexibility of lighting control in both residential and commercial spaces.

On the other hand, renters must use a track with a cord and plug for downlights, as do owners of concrete and steel structures. Track parts are highly visible. Therefore, the cord could be obscured in the corners of the wall and ceiling on its way to the baseboard receptacle, to make the installation as inconspicuous as possible. If required, a dimmer can be installed on the cord.

After most track fixtures are installed, they can be adjusted up, down, and side to side. Some are less adjustable than others. Determine the fixture limitations before you buy, and you will be able to achieve many kinds of settings.

emergency backup and dimming

Emergency backup can unobtrusively be provided by compact-fluorescent or MR-16 downlights. Such downlights accept a battery that illuminates the fixture when the electricity goes off. Clearly, this is a more aesthetic method of providing emergency lights as compared to the typical wall-mounted box and cobra-head sources. The battery can be integral within the downlight or remotely mounted nearby. The battery must be accessible and be replaced about every five years. Both residential and commercial spaces can take advantage of such emergency lighting.

Compact-fluorescent downlights with dimable electronic ballasts consume up to 25 percent less electricity than magnetic-ballast downlights. In addition, the electronic ballast allows them to be dimmed to 5 percent of light output. Further, compact-fluorescent downlights consume less electricity while dimming than incandescent dimmed to the same percent. (The light from two 18-watt compact-fluorescent sources in a downlight equals the light from a 100-watt incandescent.)

rescent is incorrect. Even though the initial cost of an incandescent downlight is less than a fluorescent, the operating cost is more forevermore.

A wall-box programmable dimmer is an excellent way to control downlights and other lights (including portable luminaires plugged into wall outlets controlled by wall switches). A programmable dimmer fits into a 3- or 4-gang junction-box space. It can create scenes. Programmable dimmers have capabilities for 3 or more scenes using 3 or more sets of dimmer slides. They can combine up to 2,000 watts of incandescent, low- and line-voltage, fluorescent (with electronic ballasts), neon, and cold-cathode sources. "Off" causes the dimmer to slowly fade to dark. Likewise, "on" causes a slow rise of light to fully illuminate. The rate of rise or fade on some programmable dimmers can be selected, usually up to 15 seconds. On other dimmers, it is set at 5 seconds. Each manufacturer incorporates different features. Compare carefully. Some features may not be required by the space. For example, in an office boardroom, a variable rate could be appropriate for changing from a speaker to video tele-conferencing. For example, in a residence, a 4-scene dimmer with 6 dimmer slides might be necessary to separately balance diverse light sources—such as $\frac{1}{4}$-watt low-voltage bookshelf lights and twin 18-watt compact fluorescent downlights. Pick and choose the features wanted. Programmable dimmers provide excellent options for delightful lighting for residential and commercial spaces. Use them and enrich the lighting choices!

COMPARE WATTS CONSUMED WHILE DIMMING

EQUAL LIGHT-PRODUCING SOURCES DIMMED TO 5%	SYSTEM WATTS CONSUMED WHILE DIMMING
100-W incandescent	30
(2) 18-W compact fluorescent	4

Clearly, the notion that using incandescent luminaires and dimming them saves over using fluo-

to change your mind about lighting

7

Life is full of changes: People change their minds, demands change around them, and their bodies change. Modifications need to be made to accommodate these changes. Lighting can be modified to accommodate some of these changes.

people change their minds

Often people like to change the style, color, and arrangement of their surroundings. Customarily, change requires a great effort, not to mention expense. However, changing lighting can be as simple as a click if the lighting fixture is attached to a track canopy. At the same time, it can be reasonably priced.

Examples

- Renters, formerly stuck with looking at a chandelier chosen by someone else, can convert for a small investment the original chandelier and a chandelier of their choice to track adapters. The original chandelier can be put away in the closet, while their chandelier graces the dining room. When the lease is up, they can put back the original and be off with no further electrician's bill and no landlord hassle.
- People who change their minds and want two chandeliers for the same room can do so. They can adapt two chandeliers to track adapters. Clip one into a track canopy one day and replace it with another chandelier the next day. For instance, hang an antique brass turn-of-the-century fixture on Tuesday and a silvery chrome chandelier on Sunday. The chandelier can be changed to satisfy mood or mind.

Instructions

A canopy, from the accessory part of track catalogs, is installed over a ceiling junction box. The canopy accepts one fixture at a time, either a track fixture made by the same manufacturer or a chandelier or pendant made by any manufacturer and fitted with the appropriate adapter. Track manufacturers have limitations for their adapters. The limitations are the components of the fixture, the diameter of the cord, the stem's thread size, or the total weight. Before purchasing, check the catalogs for the limitations of the track, and coordinate with the components of the chandelier. Track canopies and adapters are inexpensively priced. A canopy and two adapters usually cost about the same as a moderately priced restaurant dinner. Unlike a dinner, the effects of lighting are enjoyable night after night.

Put up one chandelier one day . . .

. . . and change it the next.

demands change around people

Sometimes demands require that a space to be used in a different way, either temporarily or permanently. When demands change, track components can accommodate these changes.

Examples

- When the bloom leaves the Western store business, and when hats, boots, and fringed jackets give way to an aerobic dance studio, the bright merchandising spots can be replaced with G bulbs around the mirrored walls, and they can even be wired to blink with the music.
- When dinner is to be served buffet style and the dining room table must be moved out of the way, a chandelier, formerly hung permanently over the table but now converted to a track with an adapter, can be replaced. A downlight can be put up for the duration of the buffet, and no one will bump his or her head.
- When the bridge club is coming and the dining room is to be filled with card tables, an adapted chandelier can be replaced with an adapted fluorescent ceiling fixture. The fixture will throw light broadly over the tables. After the card games, the dining table and chandelier can go back.
- When a space temporarily gets converted from one use to another use, change lighting without hard wiring. Use a low-voltage cable, rod, or rail system with low-voltage transformer either hidden in a cabinet or mounted on the wall and plugged into an outlet. Stretch cables, rods, or rails from wall to wall with 7 ft-6 in. (2.3 m) clearance from the floor (passes National Electric Code). Attach low-voltage fixtures.
- When a living space gets converted to a sleeping space, the electric track, formerly with accent lights only, can hold accent and pendant lights for different purposes.

Limitations

The possibilities of change are numerous. Many chandeliers and pendants can be track-adapted. Some track manufacturers make adapters for their own fluorescent, ceiling surface-mounted fixtures. In addition, old line-voltage fixtures can be replaced with smaller, brighter low-voltage fixtures, as long as the track and fixtures are made by the same manufacturer.

Track systems have a few limitations. Most track fixtures only fit tracks made by the same manufacturer, unless the manufacturer makes a specialty fixture, like the computerized LED fixture. (See Colored Light Sources in Chapter 20.) A track canopy can hold an unmovable track fixture.

When a relative comes to stay,
a former family room . . .

. . . can become a bedroom.

Less Light

Some people are uncomfortable in bright light. The reasons vary from vanity and self-consciousness to illness and other physical reasons that can cause the eyes to be sensitive to light. Further, if you feel that some lights are too bright, it may be neither vanity nor illness. The lighting might be poorly designed. If so, you must first determine the cause; then you can decide on the cure.

CAUSE: CONTRAST

Is there too much contrast between the light and the darkness of the rest of the space?

CURE: OTHER SOURCES

If so, add additional light sources and reduce the contrast.

See for Yourself: How Much Contrast Is There?
1. Bring several lamps into a room.
2. See how the original light appears less bright.

CAUSE: BRIGHT BULB

Is the brightness caused not by contrast, but by seeing the light bulb?

CURE 1: ADJUST

If so, and if the fixture is adjustable one, adjusted it to block the view of the bulb. Often such fixtures are installed with little regard for glare. Track lights are the most likely to offend because installers are unaware of the critical rule of thumb for track fixtures.

Rule of Thumb for Track Fixtures
Never position a track fixture to allow a person to look into it and see the bulb.

CURE 2: BLOCK BULB

If the track fixture is positioned with the bulb in view and if the fixture is not adjustable, obscure the light bulb by blocking the view to the bulb with a louver, a shade, or a baffle. Louvers are sometimes available from the manufacturers of the fixtures. Louvers obscure the bulb from many angles. Get or make a suitable louver. For example, if a table lamp blasts light up the stairs, blinding those coming down, get a perforated metal circle and attach it to the top of the shade. At other times, when obscuring devices are not available for the fixtures, they can be custom-made. The basic supplies are in do-it-yourself building supply, electrical, and hardware stores, or a sheetmetal shop can make them. However, never enclose an open downlight with a lens unless the manufacturer specifies in the catalog that it is enclosable.

Shades either block the light source completely (opaque) or diffuse it (translucent). For example, if the track bulb is bare, get "barn doors" from track-lighting accessories and hang them on the bulb. For a lamp, purchase an opaque shade that conceals the bulb.

A baffle can be made of anything that blocks the view of the fixture—fire-retardant cloth, wood, or composition boards. Arrange them to reflect light but at the same time obscure the source.

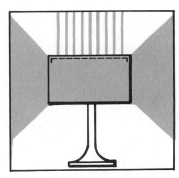

Lamp with diffuser.

CURE 3: SHORTER BULB

If the light bulb is visible, possibly it is too long for the fixture. If so, get a shorter bulb. Some identical wattages are obtainable in long or short lengths. Check with your local electric supply store. On the other hand, some fixtures are designed to have the face of the R bulb stick out. Do not install a shorter bulb in these fixtures because heat might build up inside and be a hazard.

CAUSE: TOO MUCH LIGHT

Is the brightness caused by the amount of light?

CURE: REDUCE THE LIGHT

If so, reduce the wattage of the bulb or reduce its light-producing capabilities. If it is an incandescent fixture, reduce the wattage by purchasing a lower-wattage bulb, or install a dimmer on a wall switch or on the cord. Dimmers save electricity and extend the life of the incandescent bulb by reducing the voltage. Reducing the voltage 50 percent extends the bulb life up to 16 times; reducing it 20 percent extends the bulb life up to four times. Longer bulb life and greater flexibility of settings are gained with dimmers. Both incandescent and fluorescent dimmers are available. Otherwise, a fluorescent tube can be wrapped with white, heatproof electrical tape (glass tape) found at electric supply stores for a nonvariable, permanent dimmer. The tape holds back the light given off by the tube. The more it is covered with tape, the more light it holds back. Usually, wrapping the tape with a 1-in. (2.5-cm) space between the strips is satisfactory. This method permanently dims the light. It is not as variable as a real dimmer that controls the intensity smoothly for different amounts at different times.

More Light

Perhaps the interior surfaces of a room have been changed to a dark color, absorbing more light. Or, as in the case of merchandising, a display may have been changed and the details smaller, requiring more light. Maybe more light is needed because the space is used by older people. As people grow older, they require more light, although—at whatever age—some people just need more light than others. Consequently, other light sources need to be added or higher wattages need to be used.

See for Yourself: What Is the Value of Light?
1. Take the phone book to a dim hallway and try to read the numbers.
2. Take the phone book to a lamp with a 150- or 200-watt bulb.
3. Observe how much better you can see the numbers.

CAUSE 1: DARK COLORS

Are interior surfaces—floors, wall, furniture—deep, dark colors?

CAUSE 2: OLDER EYES

Are you older than you were when you installed the lighting, or has someone older come to live with you?

CURE: FOR INCANDESCENT LIGHT

If you have incandescent fixtures, increase the amount of light by one of the following methods:

- Increase the wattage up to but not higher than the maximum specified by the manufacturer, which is usually indicated on the inside of the fixture.

- If the fixture is large enough and has the correct socket (ceramic, not brass), change at the same wattage from a reflector source to a PAR source, particularly a halogen PAR. The output will be about 18 percent more.

- Change from a flood source to a spot source, if the area of light produced is large enough. A spot source puts out about 150 percent more light but in a 60 percent smaller area. Unfortunately, there is a trade-off. Some applications will not be disadvantaged.

- Change from a line-voltage reflector or PAR to a low-voltage-type halogen beam with a line-voltage screw-base and an integral transformer (Exhibit PAR-20 and -38). The source life is 5,000 hours—over twice as long as other PAR's.

Reflector bulb.   PAR bulb.

All increases in the amount of light are relative because the eye does not perceive increases accurately. In most cases, doubling the amount of light is seen as a 50 percent increase. Likewise, increasing the amount of light and not increasing the size of the source of light concentrates the brightness. Therefore, if more light is needed but brightness is not, the amount and the size of the source must be considered. For example, the wattage of a bulb in a large Japanese lantern could be increased easily without increasing the concentrated brightness, because the large surface would disperse the light. A small lantern would not. On the other hand, a ceiling fixture previously equipped with a 60-watt bulb, but changed to a 150-watt one, would appear brighter. Whether the light would be too bright depends upon your preferences. If you want less brightness but more light, choose the largest surface to disperse the light. Excessive brightness and more light need not go hand in hand.

If you have an electric track and want more light, increase the number of track fixtures without exceeding the maximum capacity for the wattage of the track (around 1,800 watts) or for the electric circuit to which it is wired. You or an electrician can determine the capacity of the circuit.

See for Yourself: What Is the Capacity of the Electric Circuit?

1. Add up all the watts to be consumed in one circuit and divide by 120.
2. The resulting number is total amperage; it must not exceed the current capacity marked on the circuit breaker or fuse.
3. If it exceeds current capacity, the fuse will blow or the circuit breaker will trip, resulting in no electricity on that circuit.

All increases in lighting draw some electricity. However, one or two extra lights are unlikely to overload most circuits, because lighting consumes so little electricity. But if lighting is on the same circuit as small appliances - blow dryers or color televisions - it is another matter. If you are in doubt, check with an electrician.

Some elegant engineering has been accomplished in PAR sources. Halogen in PAR's has increased the amount of light produced. One company has developed a low-voltage beam in a line-voltage base with an integral transformer (PAR-38), further increasing the amount of light.

COMPARE THE PERFORMANCE
NARROW SPOTS @ 35/40 WATTS

SOURCE	BEAM	WATTS	CBCP
PAR-30 nonhalogen	10°	40	5,000
PAR-38 halogen	9°	35	7,000

CURE: FOR FLUORESCENT LIGHT

If fluorescent fixtures, use one of these methods:

- Install another fluorescent fixture elsewhere in the space.

- Install a specular reflector to get as much light as possible from the fixture.

- Install a second source; the amount increases by 85 percent. Install a third; the amount increases an additional 70 percent for a total of 155 percent more light.

- Change a T-12 or -8 fixture to a TT-5, a long compact source, and get greater illumination.

- Change a prismatic or opal lens to an acrylic or aluminum egg-crate louver, letting more direct light through. (Polystyrene louvers yellow with age.)

- Remove a parabolic louver, which pushes light down to the tabletops, and replace with a high-efficiency egg-crate louver.

COMPARISON OF PERFORMANCE
TT-5 AND T-8 @ 41°K

SOURCE	LENGTH	WATTS	LUMENS
TT-5	22 in.	39	2,840
T-8	24 in.	17	1,190
TT-5	22 in.	39	2,840
T-8	48 in.	40	2,850

At about the same length (22 or 24), TT-5's produce the highest amount of light (lumens). At about the same wattage (39 or 40), both T-8 and TT-5 produce the same amount of light. Thus, two TT-5's end to end can double the amount of light in a 4 ft (1.2 m) distance.

TT-5   T-8

If seeing better is needed, increase the amount of light as long as it is from the right direction. (See Chapter 19.) Increase the amount by using a higher wattage source or by moving the source closer. (The principle is that light decreases rapidly as the distance from the source increases. In fact, if the source is moved twice the distance away, the amount of light is decreased by 75 percent.)

Or, if seeing better is needed, increase the efficiency of the fixture by installing a reflector or by replacing a lens with a louver. Reflectors guide the distribution of the light. Louvers let more direct light through and also change the distribution. If parabolic louvers, the light is guided down to the workplane—desktop.

If less glare is wanted, shield the light source or raise the ambient light level. Also, reduce glare by avoiding a bright translucent fixture in front of a dark surface. For example, prismatic lenses on recessed fixtures appear very bright against a dark ceiling. The contrast could be glaring, especially if the ambient light is low. Reduce contrast either by increasing ambient light produced or by improving reflectance within the space with pale, not dark, surface colors.

Or, reduce glare by diffusing light. Light can be diffused by reaiming it to a large surface (the ceiling, for example). It reflects softly back into the space. Glare from direct light can be diffused by covering the source with an opal, prismatic, or polarizing lens.

Overall, whatever the lighting problem, every space should have more than one type of lighting distribution. (Small spaces that are infrequently used are the only exception!)

Rule of Thumb for Distribution in a Space

Never have only one distribution of light in a space, unless it is a small space that is infrequently used.

Change lighting either temporarily or permanently, as required; it can be done.

light
for cooks and
noncooks

8

People who like to cook need more places lighted in a kitchen than do people who do not cook. However, noncooks need minimal lighting at the work counter, at the sink, and at the cooking and eating surfaces. Most kitchens have only a center ceiling light, which gives unsatisfactory light. With a center ceiling fixture, kitchen counters are in shadow wherever you stand. Effective kitchen lighting illuminates the work being performed. Light at work locations reinforces the intended use of the location and focuses attention there. Light the kitchen by first determining the locations for lighting and by identifying where a fixture can be installed—on upper cabinets, above upper cabinets, from the ceiling, or on a shelf. Fixtures for noncooks are quickly installed; fixtures for cooks require more elaborate installation, but provide comfortable and well-integrated lighting. Some cooks may want to choose fixtures listed for noncooks, for their own reasons. After determining lighting locations, determine the amount of light needed from each fixture. The amount required varies according to the color of the kitchen surfaces; the other light sources available, including daylight; and the condition of the user's eyes. The best method of determining how much light is needed is to test it yourself.

See for Yourself: How Much Light Is Needed?

If you are considering fluorescent light:

1. Determine the length of the tube that fits.
2. Take an incandescent bulb that is at least three but not more than four times the wattage of the fluorescent tube. (For example, to test light from a 40-watt fluorescent tube, hold up a 150-watt incandescent bulb.)
3. Hold the incandescent bulb where you would install the fluorescent.
4. Observe whether the light delivered is sufficient; do not look at the bulb itself or the size of the area lighted, just the amount of light. If it is not sufficient, choose a fixture that has two fluorescent tubes in it.

If you are considering incandescent light:

1. Identify the amount of wattage the manufacturer specifies for the fixture under consideration.
2. Take an incandescent bulb of that wattage and hold it where you would install the fixture.
3. Observe whether the amount of light delivered is sufficient. If not, add more light by choosing a fixture that can deliver more, or choose additional fixtures for other locations.

At the Countertop

To illuminate work locations at a countertop, fixtures can be hung under the upper cabinet or at the ceiling, or they can be stretched between the upper cabinets. Noncooks will probably have only one counter to light; cooks will probably have several.

CABINET ABOVE THE COUNTERTOP

If there is an upper cabinet above the countertop, attach a fixture to it. Fluorescent is excellent—bright light and little heat, but low-voltage tube or strip incandescent is suitable.

LINEAR FLUORESCENT   COMPACT FLUORESCENT

Reprinted by permission of Metalux and Osram Slylvania.

If using fluorescent sources, the choices are either linear or compact. Choose lengths that equal two-thirds the length of the counter. Clearly linear is best. Linear choices are either non-hazardous which can be disposed of in the trash or hazardous which have disposal restrictions. Clearly the non-hazardous types are best. No one wants mercury in the water supply.

On the one hand, for renters, a portable fluorescent, disposable fixture with a cord and plug is ideal. On the other hand, owners can hard-wire ready-made or custom-made fixtures. Ready-made fixtures usually do not shield the source from view—not good. A miniature track with a linear fluorescent does shield. Custom-made fixtures are a fluorescent source and channel with a faceboard. Linear fluorescent can be T-12, T-8, or T-5 in 2, 3, 4, and 5 ft (6, 9, 12, and 15 cm) lengths. The 4 ft (12 cm) T-5 produces a big 104 lumens per watt, T-8 produces 89, and T-12 produces only 57. All are available in 30K (considered warm), 35K (considered neutral), and 41K (considered cool) colors of light.

Rapid-start sources provide light immediately when turned on. (Often ready-made fixtures have pre-heat sources and they flicker until they warm up.) Choose the best possible whenever possible.

- a door that hangs below the bottom shelf
- cabinet structure behind which a fixture can be placed

- a recessed lower shelf at the back that creates a place for the strip. (Inside the cabinet, the recessed shelf creates storage for short objects—canned goods, glasses, and cups.)

Low-voltage incandescent fixtures should be as long as the total length of the countertop. The tubes have sources encased in round or square plastic; the strips have bare, miniature bayonet- or double-base sources spaced along extruded metal. Either way, space the sources as close together as needed to get the amount of light desired (the closer together, the more light). However, due to heat depreciation of low-voltage sources, tube sources must be spaced at least 1″ (2.4 cm) apart. Low-voltage requires a transformer. It can be remotely mounted—on top of or inside a cabinet, or elsewhere. But, the transformer should be accessible; some codes require it. The wires are small and easy to hide, good for both remodeling and new installations. The sources burn out and should be easy to replace. Some strip fixtures have swivel sockets for convenience. (Get extra light bulbs since they are not always easy to find.) Some tubes

Put light where you work.

A door can hide the fixture.

A recessed shelf holds a fluorescent light.

have to be sent back to the factory for replacement. Some low-voltage fixtures have a shield with their slim profile; others need shielding. Some must be cut by the factory. Others can be cut on the job. Sources are available in 1 to 5 watts. Like fluorescent, install low-voltage at the front of the upper cabinets when possible. Install at the back only if the backsplash is a matte finish. Glossy backsplashes reflect like mirrors any sources hung on upper cabinets.

Sometimes a diffuser or louver is needed with a fluorescent strip. The cabinet needs to be 3½ in. (8.9 cm) deep to hold a diffuser or a louver. A diffuser spreads the light and reaches a large area. It can be prismatic or opal. A louver controls light in case the tube might be seen by someone seated in the same room or in an adjacent room. A louver can be an egg-crate grid, a parabolic wedge, or some other type.

It is not always necessary to have custom-made cabinets to accommodate fluorescent strips. Some ready-made cabinets can incorporate them with minor modifications. However, modify them before hanging on the wall; afterward it is more difficult.

NO CABINET ABOVE THE COUNTERTOP

When there is no cabinet above the countertop, illumination can come from a recessed, surface-mounted, or suspended ceiling fixture. A re-

cessed fixture is the least obtrusive. If possible, choose a fixture with a lens or diffuser. A lens either spreads or concentrates the light; a diffuser scatters the light and obscures the source. Both keep the inside of the fixture clean, sustaining the amount of light produced. Do not use an open ceiling fixture—a downlight with no lens or diffuser—unless extra brilliance is required. In such circumstances, higher wattage is needed. An open fixture gets dirty quickly, and if it is used alone, the light can be harsh. Ceiling fixtures are far away from the surface to be illuminated and need to pump more light. Use a source that will provide 2,000 or more lumens—a 32-watt (or 42-watt) compact or a 48-in. (1.2 m) linear fluorescent.

Also, if there is no cabinet above the counter, illumination can come from a pendant or a suspended fixture. The fixture can be decorative, and it can provide a repeatable interior design feature if coordinated with the same fixture elsewhere. Hang a pendant high enough to allow for headroom. Choose one that transmits light through the sides as well as down. Otherwise, the light might be too harsh if it is the only source.

Finally, if there are two side cabinets, illumination can come from a lighted shelf stretched between them. The shelf produces light and holds kitchen items. The light can go up and down, spreading over more than just the countertop. Three benefits are received from one shelf. The shelf can be made of wood, glass, or plastic. Use a fluorescent strip as a light source, and get one as wide as the shelf. Mount it under the front edge. If possible, use a rapid-start fluorescent 36-in. (91-cm) or 48-in. (122-cm) strip. It turns on instantly and is, therefore, the most suitable.

At the Sink

The most frequently used place in the kitchen is the sink. Light it well. The amount of light should be enough to help distinguish between a dirty dish and a clean one. The fixture can be attached to the upper cabinet or to the ceiling. If there is a cornice board between two upper cabinets at the ceiling, a fixture can be mounted behind it.

ON THE CEILING MOUNTED BEHIND A CORNICE BOARD

If a fluorescent source is to be used, mount it behind a cornice board. (See Built-In Cornice Light in Chapter 5.) A linear fluorescent should be as long as possible. Choose a non-hazardous tube that provides at least 2,000 lumens, or more if the surface colors are dark, utilizing the high-output fluorescent when necessary. A compact-fluorescent source should be 32 or 42 watts. Choose an electronic ballast for the best energy consumption and flicker-free operation. Some are dimmable. The light is shadow-free without highlights. The color of the light should blend with the other sources in the room. Choose 30K if all the other sources are non-halogen incandescent. Choose 35K if some are fluorescent and some incandescent. Choose 41K, if the kitchen is all white.

If an incandescent source is to be used, choose infrared type, either line-voltage, 100-watt PAR-38 or low-voltage, 75-watt MR-16. These sources consume more electricity than fluorescent, but they produce more light. The light is perceived as bright with shadows and highlights. However, they will put heat into an already hot kitchen—unwelcomed in some climates.

Initially, the most inexpensive source and fixture would be a 100-watt halogen A (BT, TB, or MB) in a porcelain socket. Ultimately, it would be the more expensive.

MOUNTED WITHOUT A CORNICE BOARD

If a fluorescent source is to be used, choose a 32- or 42-watt compact fluorescent source in a recessed or surface-mounted fixture. An open, high-

Lighted shelves also hold kitchen items.

quality recessed fixture with a 42-watt source produces 35 footcandles (350 lux) on the countertop and a lensed fixture produces 30 footcandles (300 lux).

If an incandescent source is to be used, choose infrared-enhanced sources—100-watt, PAR-38 line-voltage, or 75-watt MR-16 low-voltage. Incandescent fixtures adapt to sloped ceilings and can be aimed. With ceilings over 9 ft (2.7 m), consider using small apertures to reduce glare from the lighted source. Pinhole or slot apertures conceal the best. If several are to be used, pinholes are the best choice. Avoid using more than two slot apertures on one ceiling.

Choose the beam width according to the size of the sink. Consult track or low-voltage fixture catalogs for beam-spread charts. Measure the distance from the ceiling to the countertop. Unfortunately, distances on the charts are not always the exact distance needed. Round the numbers and then decide if more or less light is needed—usually more is good at the sink. For a flat ceiling, read the chart at the 0° aiming angle at the distance needed. Determine the footcandles expected. It should be at least 30 and could go up to 50 depending upon reflectance of surfaces receiving the light and age of user. (Dark surfaces and older age demand more light.) Likewise, determine width of beam. It should be as wide as the sink. Note example; it shows that a 20-watt MR-16 would give 18 foot-candles and a 42-watt would give 40. Err on the side of more rather than less at sinks, because light sources produce less light as they age. Surfaces get dirty and reflect less. Consequently, provide a little more at the beginning.

LAMP	BEAM SPREAD (To 50% Max. CP)	MAX C.P. (Candelas)	RATED LIFE (Hours)	0° AIMING ANGLE (A)			
				DIST-ANCE (D)	F.C.	BEAM LGTH (L)	BEAM WIDTH (W)
20W MR-16 FL (T-H) (BAB)	36° × 37°	460	2,000	2'	115	1.3'	1.3'
				3'	51	1.9'	2.0'
				4'	29	2.6'	2.7'
				5'	18	3.2'	3.3'
42W MR-16 FL (T-H) (EYP)	36°	991	2,500	3'	110	1.9'	1.9'
				5'	40	3.2'	3.2'
				7'	20	4.5'	4.5'
				9'	12	5.8'	5.8'
50W MR-16 FL (T-H) (EXN)	37° × 39°	1,500	3,000	3'	167	2.0'	2.1'
				5'	60	3.3'	3.5'
				7'	31	4.7'	5.0'
				9'	19	6.0'	6.4'
75W MR-16 FL (T-H) (EYC)	38° × 40°	2,000	3,500	4'	125	2.8'	2.9'
				6'	56	4.1'	4.4'
				8'	31	5.5'	5.8'
				10'	20	6.9'	7.3'

Reprinted by permission of Lightolier.

If a decorative accent is to be created or continued, a ceiling-hung pendant or two wall-hung uplights on either side of the sink could be used. A screw-based compact fluorescent source could be installed in the pendant for energy-efficiency and the light, having less distance to travel, will be brighter than if originating from the ceiling. If uplights are the choice, make sure the surfaces receiving the light are pale in color and can reflect as much light as possible. Also, install the uplights at least 10" (25.4 cm) down from the ceiling for good light distribution.

At the Cooking Surface

Even noncooks need to see when the water boils. Cooks need to see a lot more. Consequently, cooking surfaces (ranges, stovetops, or built-in units) need light. If a hood over the cooking surface has a light, use it with the wattage specified by the manufacturer. Do not reduce the wattage unless it creates a glare. A screw-base compact fluorescent source could be substituted if space permits and energy conservation is the goal. If there is no hood, install a shielded fixture—shielded both from view and from splatter.

If upper cabinets are available, hang a well-shielded fluorescent fixture or build in an incandescent low-voltage or line-voltage fixture with a lens. Almost all low-voltage tube or strip lights are unsuitable to withstand the high temperatures of cooking surfaces.

If upper cabinets are not available, a lighted shelf, a pendant, or a ceiling fixture can illuminate the cooking surface. (See At the Sink.) The amount and color of light is more critical for cooks who need to distinguish between opaque and translucent while stir frying. Noncooks need to distinguish between browned and burned. But both cooks and noncooks will want at least 20 footcandles—usually the light provided from 30 fluorescent tubular watts or 65 to 75 watts line-voltage incandescent. Use fluorescent light whenever possible; it is cooler, and the cooking surface produces enough heat. Special infrared incandescent bulbs or warm-hold microwave ovens can keep food hot elsewhere. Save energy and minimize heat that could demand more air-conditioning.

At the Eating Surface

An eating surface may be a counter, a bar, a peninsula, or a table. Noncooks are more likely to eat in the kitchen than are cooks. Eating requires illumination, unless you like eating in the dark. Eating surfaces thrive on light. The more direct and intense it is, the more appealing the food looks. Incandescent bulbs produce the most direct and intense light.

Direct and intense light can come from any point-source fixture that puts light down hard on the eating surface—open downlights, either surface mounted or recessed, as pendants, or in chandeliers. (Currently, residences utilize incandescent point sources, rather than metal-halide or white-sodium point sources. But that will probably change.) Low-voltage, incandescent point sources (MR or PAR) give the greatest intensity for the lowest wattage. Line-voltage PAR point sources can give intensity at lower wattage than line-voltage R (reflector) sources. But, even a well-engineered A-source downlight can give sufficient intensity. (Check manufacturers' catalogs for data on fixtures with these sources. See Interpreted Calculations in Chapter 21.) All these sources enhance the look of food and tableware with sparkle, shadow, and vibrant color. Make every meal a visually appealing experience.

FRAMING PROJECTOR

The most distinctive and intense light for eating surfaces is produced by a framing projector. In theaters it is used to light stage objects precisely. In eating areas, it can illuminate just the eating surface and nothing else. It is capable of cutting the edges of a light beam. When fully open, it can light an area about 60 in. by 60 in. (152 cm by 152 cm). When closed down, it can light a smaller area. The light can be confined to shine brilliantly on all table settings and food, enhancing colors and appetites. The light source is a tungsten-halogen incandescent bulb. It requires careful handling, but burns brighter longer than other lights. The fixture needs readjusting each time the bulb is replaced. These requirements are fine for some people and terrible for others. The framing projector can be either recessed in the ceiling or track-mounted. Thus, both owners and renters both have equal opportunity for using this precise optical instrument.

CANOPY

Fluorescent lights can be suspended over the eating surface, either in ready-made fixtures or built into a canopy. The ready-made fixtures are slick-looking and accent any nontraditional interior. A canopy is a wooden frame, open above and covered with a diffuser below. The light from the fluorescent strips bounces off the ceiling and down through the diffuser. Make the canopy proportional to your eating surface and within the sizes of fluorescent strips (2 ft to 8 ft, or 0.6 m to 2.4 m). Suspend it 10 in. (25 cm) from the ceiling in an average-height room, lower in a higher ceilinged room. Make the depth of the canopy at least 6 in. (15 cm). The distance between the strips should be one and a half times the depth of the frame. The distance from a strip to the side of the frame should be half the distance between the strips. Therefore, if the frame is 6 in. (15 cm) deep, the strips should be 9 in. (23 cm) apart.

Energy Conservation

Some states are determined to reduce energy consumption. California, for instance, requires that the ambient or general lighting in kitchens (and baths) be energy efficient—the source must produce at least 25 lumens per watt. Fluorescent sources meet this requirement. Incan-

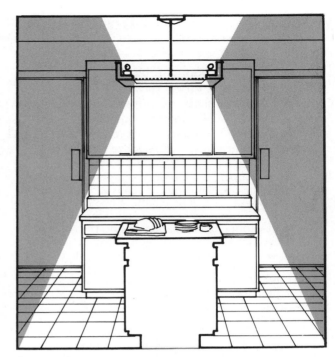
A canopy bounces light up and down.

descent sources do not. But, only tubular or other plug-in fluorescent are allowable. Screw-based compact fluorescent are not permitted. They are too easily interchangeable for energy-inefficient incandescent after the electrical inspection is completed. In addition, the fluorescent sources must be connected to first accessible wall switch. Other sources (fluorescent or incandescent) for localized lighting can be connected to second or third switches. Other state (or national codes) are based on maximum allowable watts per square foot of the space. They usually apply to laundry rooms, too. Some states regulate minimum efficiencies acceptable for fixtures. These methods, of course, get at the same problem—wasted energy.

Local building inspector's office can supply information on requirements, including special fixtures for closets, damp locations and other potentially hazardous locations.

The whole kitchen can be pleasingly lighted with fluorescent. Choose warm-white deluxe or prime-color sources. They render colors well and blend with incandescent light in other spaces. Prime color costs about three times more but is worth considering. Kitchen surfaces are used 365 days each year and the additional cost would be less than a half a cent a day for the life of the source. Sell the long-term benefits.

light for the room itself

Sometimes the room can be illuminated sufficiently by the light from the sink, the stove, and the other work locations, but sometimes it cannot. Cooks will want to see easily into upper and lower cabinets, and they may require more light sources. Several alternatives are available.

- If the upper cabinets do not go all the way to the ceiling, cove lighting can be installed on cabinet tops with excellent results. Cove lighting consists of fluorescent strips that bounce light off the ceiling and spread it around the room. Collectibles or usable kitchen objects can be displayed above the cabinets and lighted either from in front or from behind. Choose whichever lighting looks best and installs easiest. (See the directions in Chapter 9.)
- Fluorescent lights can be built into a dropped soffit above and in front of the upper cabinets, or they can be built as wall brackets in the same position. (See the directions in Chapter 9.)

Cabinets support cove lights.

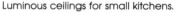

Luminous ceilings for small kitchens.

- A center ceiling fixture can be either surface-mounted or recessed. A recessed fixture with a dropped lens spreads the light on the ceiling; a flat lens sends the light down. Some surface-mounted fixtures permit the light to come through the sides of the fixture; some do not.
- If the fixture can be seen from another room, choose one that does not send light through the sides. When used alone, too much glare would be created, because the light would contrast with the darkness of the room.
- A center ceiling fixture illuminates the center of the room, which in most kitchens is empty. Therefore, the light should not be overly bright and attract attention. Work locations should be brighter. Countertops always look better than floors. If the center is not empty and contains an island or a peninsula, install the ceiling fixture directly above it.
- In small windowless kitchens, luminous ceilings are excellent. They provide broadly spread room lighting from low-energy-consuming fluorescent strips. They must be augmented by undercabinet lights; otherwise, the cook creates his or her own shadow. (See the directions in Chapter 9.)

other lighting

Other lighting is available for kitchens.

- A warming light, an infrared reflector bulb, can keep food warm before serving. It can be built into an upper cabinet. These bulbs are particularly useful at passthroughs. A 125-watt infrared bulb is sufficient. It heats up and cools down fast. The heat it produces is clean and concentrated.
- Lighted cabinets with glass or plastic cabinet doors transmit the light. (See directions for furniture lighting in Chapter 11.)
- Luminous panels can be used between the countertop and the upper cabinet as the backsplash or between two upper cabinets as a cornice. Both provide not only light but also a decorative ambience. The panels can be glass or plastic. For the backsplash position, choose panel colors that do not impart a distinct hue to the light. Red, green, or blue, for example, would make food look odd. For the cornice position, choose any hue. The light below will still be white, and the light transmitted through the panel will be colorful without causing interference. In the backsplash position, a panel should be illuminated by fluorescent strips. Build a box for them. Place them at the top and the bottom, or at the left and right sides, 4 in. (10 cm) back from the edges so that they will not show. To obtain even light, the box should be one-sixth as deep as the distance between the strips. For example, if the strips are 21 in. (53 cm) apart, the box depth could be 3½ in. (8.9 cm). Paint the inside of the box flat white. Hinge the panel so that the tubes can be replaced when they burn out, usually not sooner than three

years. Moreover, plan how you want to control the luminous panel; a wall switch is the most convenient to use.

A stained-glass luminous panel between upper cabinets.

programmable dimmers

All of the fixtures in a kitchen do not need to be on at the same time. However, I had one client who insisted that all the kitchen lights be on one switch, all on or all off. She did not want to switch lights at different locations, no matter how convenient it was. Contrary to this feeling, the advantage of multiple lighting is that lights can be used alone or in combination, where and when needed. Multiple lighting is energy-efficient, since unused light can be turned off.

Programmable Dimmers

Programmable dimmers can control the light of multiple fixtures in the kitchen (or in any room). The program combines the light levels of several fixtures on different dimmers at one on/off switch. Each dimmer can control up to 2,000 watts. The program's memory stores the combinations. Programmable dimmers have 3 or 4 switches, hence can create 3 or 4 different scenes. One scene for a kitchen could be low, soft light to permit finding the ice and hors d'oeuvres. Another scene could be all sources on brightly for serious pot scrubbing. A third scene could be bright light focused on the eating surface and soft light at countertops where pots and pans go unnoticed. These dimmers work well for open kitchens or for kitchens that host frequent parties. Light up your life in the kitchen—easily and satisfactorily.

light
to dress and
bathe by

9

To clothe and bathe ourselves we need mirrors, tubs or showers, and closets. Sometimes we need ironing boards and exercise areas. All of these require light. Light for mirrors must be on the person, not the mirror. Light for tubs, show-ers, and closets must be in the tubs, showers, or closets. Light for exercise areas should not be on the person, but should be reflected from some-where else in the room.

at mirrors

In the Bedroom

Mirrors above a dresser or dressing table are commonly lighted by lamps. The design of the lamp and shade is not just a matter of aesthetics but of function. If the shade is opaque, narrow at the top, and broad at the bottom, it squeezes the light down, illuminating the top of the dresser, not your face. Choose a shade that transmits light and is 2 in. (5 cm) taller than the bulb. If the lamp is tall enough, it will illuminate your face well; if it is too short, glare gets into your eyes. It almost cannot be too tall, unless the bottom of the shade is above your eye level.

Rule of Thumb for Lamps at a Mirror
Place a pair of lamps about 3 ft (0.9 meters) apart on either side of the mirror.

Experiment with various size bulbs to get the exact amount of light desired. Instead of an in-candescent bulb, consider using a fluorescent adapter, which holds a circular fluorescent tube and adapts to the screw-base socket, or use the newer bulb-shaped fluorescent sources, which have excellent light output and a screw-base. Fluorescent sources emit more than three times as much light as incandescent and use at the least one-third less electricity.

Other bedroom mirror light sources could be wall-hung, ceiling-recessed or surface-mounted fixtures, two pendants, or a chan-delier. (Refer to Chapter 3.)

A Full-Length Mirror

Ideally, the fixture that lights a full-length mir-ror should be out of sight but should illuminate the person fully. This is very hard to do. On the ceiling, position a fixture close to the wall, and include a diffuser to soften shadows under the eyebrows, nose, and chin. Usually full-length mirrors are located on doors or short walls where ceiling fixtures fit, both architecturally and aesthetically. Architecturally, they are out of the way; aesthetically, they do not attract unnecessary attention to themselves. On the wall, position one fixture above the mirror or two on the sides.

In the Bathroom

The usual location for a bathroom mirror is on the wall above the sink. The size of the mirror determines which lighting fixtures should be used, and the status of ownership determines how the fixtures can be installed. Owners can more easily install fixtures connected to the electricity in the wall or ceiling. Renters, on the other hand, must use plug-in wall fixtures. Either way, the fixtures should control the brightness and distribute the light.

SMALL MIRRORS

Lighting small mirrors is difficult because the light should be distributed along the side of the face, the top of the head, and under the chin. Hard to do! Any bright source of light within view—along the sides of the mirror—causes the eyes to respond to brilliance while trying to see a more dimly lighted, highly detailed image in the mirror. The best solution is linear fluorescent, spreading light broadly from beyond the cone of vision—55 degrees up from vertical. There are many ways to do it.

A dropped canopy reflects light from many surfaces to the face.

Light on both sides and above the mirror reflects on your face and neck.

- *A dropped canopy.* Construct and hang from the ceiling a 2- or 4-source fluorescent canopy that distributes light below and above. Hang it 10 in. (25.4 cm) down from the ceiling. Use a lens below and leave open above. The light will bounce off the ceiling, the walls and countertop (if there is one) or the surface of the pedestal lavatory. The fluorescent need not be cool white. It can be a wide range of color choices that enhance facial colors and distinguish beard from skin.
- *On the wall at each side and/or above the mirror.* Use incandescent or fluorescent. But make sure the source is covered with opal glass or other diffusing lens. Do not use bare bulbs, unless low wattage and placed well out of direct view to the right or the left. Fluorescent sources are preferable because they provide good distribution for little energy—20 watts on each side and 80 watts above for a total of 120 watts and 15 percent extra for the ballast. Incandescent would require three times as many watts and therefore electricity to do the same job. Also, incandescent produces heat.
- *Bare bulbs above the mirror.* Bare bulbs can be used above the mirror if they are low wattage.

For a dramatic effect,
fluorescent light behind a mirror.

look appropriate with large built-in lighting. Fluorescent fixtures are large sources of light, can be surface-mounted, and are architecturally suited to being built in. Owners may want to build a dropped soffit or a lighted wall bracket to hold fluorescent tubes. A soffit directs light down, reflecting it from the countertop and the mirror to the face. Likewise, the bracket directs it down, and it also directs it up if desired to flood the ceiling with light. A ready-made surface-mounted ceiling fixture can be used by owners who choose not to build in lighting or by renters who cannot build in. A ceiling fixture over the mirror provides light in the same way as a soffit or bracket. Renters can connect them to a track hooked up to the electricity through a ceiling junction box where a ceiling fixture is now located.

- *Dropped soffit.* A dropped soffit is a boxed-in area of a ceiling dropped 8 to 12 in. (20 to 30 cm) down. Dropped soffits look best above a vanity top and should be the same length (left to right), but not the same depth (front to back) as the vanity. The soffit can be 12 to 18 in. (30 to 46 cm) deep and should be able to hold a diffuser to scatter the light. Use two singles or a double fluorescent strip, unless the room has medium- to dark-colored walls or unless your eyes require more light. If so, use three or four fluorescent strips. Paint the inside of the soffit flat white to reflect as much light as possible. The vertical surface of the soffit should be finished to blend with the bathroom walls, and the bottom should blend with the ceiling. For example, a soffit could be built of sheetrock, plastered, painted off-white on the bottom to match the ceiling, and wallpapered on the outside to match the walls.

Coated bulbs produce the least glare. If clear bulbs are preferred, use other light sources in the space to dilute the glare. Low-wattage, line-voltage G bulbs in 15 watts or less are suitable. But, low-wattage, low-voltage G bulbs are better. They come in 2½ to 5 watts, and large or small sizes. Low-voltage must be hooked up to a transformer, however. Mount it in the cabinetry or the attic above. In general, many low-wattage bulbs mean less glare. Few high-wattage bulbs mean more glare.

- *Behind the mirror.* For a dramatic and decorative effect in guest bathrooms, both residential and commercial, where mirrors are used casually rather than seriously, fluorescent or tubular incandescent behind the mirror are unexcelled. Mount the mirror on a wooden frame 4 in. (10 cm) smaller than the dimensions of the mirror and 4 in. (10 cm) deep. Attach the strips to the frame, at the top, the bottom, and the two sides. The mirror will be surrounded by a halo of light and appear to float. This effect is particularly flattering for dressing rooms that are intended to thrill the customer with his or her image.

LARGE MIRRORS

If a mirror is 3 ft (0.9 m) wide or more, it helps disperse the light by reflection. Large mirrors

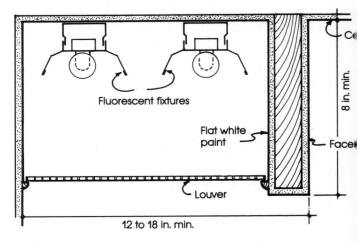

Fluorescent fixtures

Flat white paint

Louver

Ceiling

8 in. min.

Facer

12 to 18 in. min.

Soffit details.

A dropped soffit puts light down.

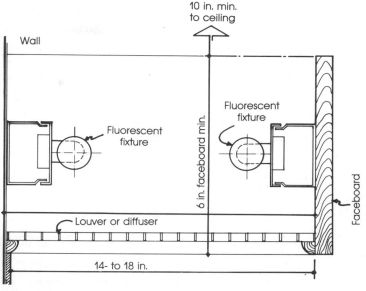

Bracket details.

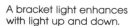

A bracket light enhances with light up and down.

- *Bracket.* A bracket is a cantilevered shelf holding lights. Like a soffit, it looks best above a vanity top. It should be the same length as the top, be at least 10 in. (25 cm) down from the ceiling, to allow adequate reflection, and be 14 to 18 in. (36 to 46 cm) deep. The face board should be at least 6 in. (15 cm) high and covered with any material that enhances the interior decor—wood, tiles, wallpaper, fabric, or paint. Use two single fluorescent tubes to light the bracket, one on the inside of the faceboard at the top and one on the wall. The distance from the end of the strip to the end of the bracket, ideally, should be 6 in. (15 cm), in order to have smooth lighting. Paint the inside of the faceboard flat white. Use a diffuser at the bottom. Use white perforated metal on the top, covering just the fluorescent tube at the wall or covering both tubes for less light on the ceiling.

Individual differences, such as individual preferences, eye conditions, and interior colors, play a big part in choosing the type and amount of lighting for mirrors. Some people prefer theater lights beside the mirror. Some have preferences about the color of the light. Incandescent light enhances colors in all faces. Fluorescent can do so also, if warm-white deluxe or prime-color tubes are used. Such tubes produce more red,

reflecting a truer skin color than do cool-white tubes. However, people who work in offices may want cool-white to apply makeup, since their office light is the same color. (Some portable makeup mirrors are made to light both ways.) Also, older users require more light—between two-thirds and twice more than others. Consequently, more wattage is required.

Different interior colors also affect the color of the light. To this point, one South Florida resident wrote to my newspaper column inquiring why her skin tone appeared different in two bathrooms equipped with the same kind of theater lights at the mirrors. In one room her skin tone looked true; in the other her skin appeared sallow. The reason was the color of the walls—a sunshine yellow—reflecting an unflattering hue on her. Never cover a large area in any space with a color that does not flatter you. The light, whatever the source, will pick up a tint of that color and, as a result, reflect that color on you.

protective wall receptacles near sinks

Wall receptacles near sinks should be the protective type—ground-fault circuit interrupters. They guard against hazards of shocks originating in electrical appliances such as hair dryers and toothbrushes. (Shocks from the electric wiring in the wall will be controlled at the circuit breaker in the main power panel.) A ground-fault interrupter will shut off the electricity quickly when the flow is interrupted, even just a bit, preventing a shock. Electricity can be interrupted by faulty or worn-out appliance wiring or by getting the wires wet. These protective wall receptacles are available at local electric supply stores in a single or double size. They fit into the same space as the old receptacle. In many areas, building codes require them for new construction when a receptacle is located

Protective wall receptacle.

within 8 ft (2.4 m) of a kitchen or bathroom sink. People would do well to install them in older homes, particularly in bathrooms where teenagers use electrical appliances.

in closets

Light is mandatory in closets for inspecting, selecting, and finding clothes. Consequently, all clothes closets should have a light. The National Electric Code requires a lensed, not open, fixture for closets. Choose from one of these:

- A lensed recessed downlight with a 32-watt compact fluorescent source or a 75-watt A source, preferably halogen—BT, TB, or MB. (A 40-watt A source or 13-watt compact fluorescent cannot provide enough light to distinguish colors.)

- A recessed or surface-mounted, enclosed, linear fluorescent two-thirds the width of the closet with

a high color-rendering T-8 source. The choice is an 85 or a 95 (CRI). Use the 30K if mirrors are in the closet.

Closet lights can be controlled by a wall, automatic door switch, motion detector for large closets, or a timer switch for those who leave the closet door open and the light on too long. A door switch operates with standard swinging, not bifold or sliding, doors.

If closets have built-in dressers with mirrors, light them with fluorescent sources. Backlighting is particularly flattering.

at the ironing board

Many of my lighting clients claim that they do not own an iron. Nevertheless, for those who do, ironing needs strong directional light to reveal wrinkles—incandescent, not fluorescent. Just being able to see the fabric reasonably well is not sufficient. Often clothes are ironed under low indirect light, and the wrinkles remain. They appear later in brighter directional light—on the way to the car, or outside while waiting for a taxi. Then it is too late.

Light from an oblique angle can reveal the wrinkles while the iron is in hand. To light the board minimally, use a 75-watt reflector bulb, either in the ceiling or in an adjustable floor lamp, and a fixture elsewhere in the room. Use higher wattages if you iron more than minimally, such as ironing what you sew.

over tubs and showers

Tubs and showers require a fixture and usually do not have one. The fixture must be watertight (lensed) and should be recessed in the ceiling. It should be able to produce 8 or more footcandles from an 8-ft (2.4 m) ceiling to be able to see in a tub or shower. Most do not!

Sometimes a tub or shower is a special bathroom feature. An owner with a bathroom accessible to the roof can install a skylight to illuminate during the day and reveal the sky at night. (See Chapter 17.) A skylight can be enhanced with low-voltage, low-wattage, watertight tube lighting of $\frac{1}{4}$ to 1 watt size. The sparkle is delightful.

A luminous ceiling can substitute for a skylight. It provides a large surface of light. A fluorescent-strip luminous-ceiling or remote-source system can provide the light.

A fluorescent-strip luminous ceiling consists of linear fluorescent sources spaced evenly behind translucent diffusing panels in a grid ceiling typically 6 in. (15 cm) deep. The deeper the luminous box, the more uniform the brightness of the light on the panels. Whatever the depth, position the sources apart at a distance equal to one and a half times the depth of the box. Allow half that distance at the edges of the box. Thus, if the box is 6 in. (15 cm) deep, the fluorescent sources should be 9 in. (23 cm) apart and 4½ in. (11 mm) in from the edge on each side.

If the tub or shower area is embellished with plants, such a luminous ceiling would support the plants if kept on for 14 hours per day and equipped with 41K sources.

Sample Electric Cost

One 32-watt 4 ft fluorescent source, including the ballast, can be on for 14 hours per day for $1.20 per month at 10¢ per kilowatt-hour.

A remote-source luminous ceiling consists of an illuminator box with either a metal-halide or MR-16 source and prismatic film in the grid as the diffuser. The light will be uniform and cool.

in exercise areas

Some people exercise at home and some exercise in commercial spaces. Exercise spaces need general room illumination, not lights at specific locations. One form of general illumination is a ceiling fixture that distributes the light, and is itself not overbright. These qualities are controlled by fixture design and diffuser. If the fixture design permits the light to spill through the sides as well as down, it spreads the light. If the diffuser is or poor quality, it will be too bright and the light source will be visible through it. Opal or ceramic-enameled diffusers are the best for incandescent fixtures. But if the fixture is small and overwatted, even a good diffuser can not overcome the annoying glare.

For fluorescent fixtures, diffusers that are

A tub with a luminous ceiling keeps plants alive.

rated to have a high visual comfort probability (70 or more) are the best. Fixture manufacturers indicate this rating as VCP. The better the VCP rating, the more comfortable the lighted diffuser appears. Comfort is always a lighting objective.

Many exercises are done lying on one's back, facing the ceiling. Therefore, do not position the fixtures directly over an exercise bench or mat location. If mirrors are included in the exercise space, be aware that the fixture will reflect in the mirror. The most glare-free lighting for an exercise space with mirrors is cove lighting. It is indirect. It is composed of a cove board on two or more walls, with fluorescent tubes behind it. Install the cove board at least 10 in. (25 cm) down from the ceiling to bounce as much light off the ceiling as possible. The ceiling then becomes a large light source softly reflecting in the mirrors, not glaring. This technique is particularly suitable for commercial exercise spaces, where the patron's degree of comfort is critical for business.

Within limits, the farther down the wall the cove is installed, the broader the light is distributed across the ceiling and the softer the brightness. The ceiling (and the wall) above the cove should not be too bright. It should not exceed 500 footcandles of reflected light, otherwise it will attract too much attention and destroy the visual balance. Close and even spacing of light sources behind the faceboard is critical for a smooth wash of light on the upper wall.

To spread the light well across the ceiling, use asymmetric reflectors or install the fixtures on a block angled at 20 degrees. The problem with cove lighting is that the light gets duller as dust accumulates in the cove. Coves should be dusted every time the room is.

Rule of Thumb for Position of Cove Lighting on the Wall

Coves on one side of a room:	position down $\frac{1}{4} \times$ ceiling width
Coves on two sides of a room:	position down $\frac{1}{6} \times$ ceiling width

---

Fluorescent, tubular-incandescent, and cold-cathode sources can be used in coves. Fluorescent is excellent, because it gives more light for the watts consumed, does not put out much heat, and is available in incandescent-like color of light. Tubular-incandescent sources are the least desirable due to the high consumption of electricity and the heat generated. Cold cathode does not produce as much light as fluorescent,

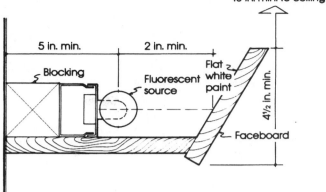

Flat ceiling at 10 in. down.

but can yield a continuous line of light. Fluorescent sources can overlap or highly engineered reflectors can smooth the light.

Some ready-made fixtures can be tilted up to 20 degrees to aim the light. The fixture has three angles to choose from. Custom-made coves also can have 20 degrees blocking for the same purpose. If the ceiling is flat and the cove is close to the ceiling (10 in. or 25 cm or less), use angled blocking and angled faceboard. If the ceiling is vaulted or sloped, or the ceiling is flat and the cove is far from the ceiling (18 in. 46 cm), use a vertical faceboard.

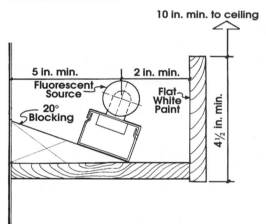

Sloped or vaulted ceiling, or far from ceiling.

Indirect Remote Source Lighting

Lighting spaces indirectly with a remote source is possible, thanks to total-internal-reflection technology. Such technology uses light guides of hollow prismatic film or side-lit fiber optics with a single source at one end. (With long coves, a second source at the other end could be needed). The light guides are essentially fixtures of highly engineered materials that distribute the light well by reflecting and re-reflecting the beams inside the guide. (See Remote Source Lighting in Chapter 20.)

An example of internal reflection occurs when you see a reflection of a fish in a fish tank at the same time as you see the actual fish. The second fish is created by total internal reflection.

See for Yourself: Total Internal Reflection

1. Stand in front of a rectangular fish tank that is free-standing (has space behind it).
2. Look at a fish near one side of the tank. You should notice that there appears to be 2 identical fish. (The interior of the tank and the fish are reflected on the side of the tank and then reflected into your eyes.)
3. Walk around to the other side of the tank. (The illusionary second fish disappears.)

Adapted with permission from General Electric.

Light guides for coves use either an MR-16, a metal-halide, or a sulfur fusion source.

The source type depends upon the amount of illumination needed. Use a fusion source for very high ceilings; use xenon metal-halide for other heights and possibly avoid the necessity for a fan in the illuminator box.

Rule of Thumb for Determining Lighting System Efficacy

$$\text{Efficacy} = \frac{\text{watts into a fixture}}{\text{lumens out}}$$

Like all lighting fixtures, the system (source and light guide) can be judged by lumens per watt (efficacy) and by efficiency. Typically, efficacy of fibers is 5 to 10 lumens per watt and efficiency is between 9 and 18 percent. Choose the combination that yields the highest persent.

Rule of Thumb for Determining Lighting System Efficiency

$$\text{Efficiency} = \frac{\text{lumens into a fixture}}{\text{lumens out}}$$

Hollow Prismatic Light Guides

The hollow light-guide system consists of a remote illuminator box with the light source, a 3 to 8 in. (8 to 20 cm) hollow square of prismatic film, a diffusing surface inside reflecting the light toward the ceiling, and mirrors at right-angle bends and the opposite end from the source, if a second source is not needed.

Ready-made light guides are 8 in. (20 cm) and 40 ft (12 m) long. Two guides can be linked together for 80 ft (24 m) of continuous light for straight runs. Other configuration and sizes must be custom-made.

Fiber Optic Light Guides

The side-lit fiber-optic system consists of a remote illuminator box with a light source and a side-lit fiber. The fibers are either glass or plastic.

The glass fibers can transmit light the farthest with the least loss of light. At 99 ft (30 m) these fibers transmit 74 percent of the light, while other fibers can be as low as 7 percent. And they last longer.

Fibers cannot make right-angle bends. Bend radius is critical to preserving the total internal reflection and manufacturers specify permissible radii.

Rule of Thumb for Fiber-Optic Bend
Radius of the bend = greater than 10 x fiber diameter

---

accent your paintings with low voltage

10

A painting is a special purchase. It is expensive and is proudly displayed on the wall, at home or at the office. It is visible, but yet not visible. At night, like the rest of the walls, the painting fades away forgotten. Turning on lamps does not help. The painting needs its own light.

Paintings can be accented with line-voltage or low-voltage sources. The choice depends upon how much light is needed, how much light can be tolerated by the artwork, and the beam spread required.

First, how much light is needed? Dark spaces and dark paintings require more light than pale spaces and pale paintings. Dark spaces might need line-voltage sources or high-wattage low-voltage sources. At the same time, be certain that the artwork can tolerate the amount of light intended without fading or be certain that the owner is tolerant of the fading. (Many owners want the light and do not care about preserving art for the future.) However, if fading is a concern, the Smithsonian Institution has established rules to follow. The maximums recommended for exposing art are:

15 fc for watercolor paintings and fugitive dyes
20 fc for stable dyes
30 fc for oil paintings
40 fc for painted wood or fabric and stable dyes.

Rule of Thumb for Art Fading
Do not expose art that can fade or deteriorate to more than 5 fc for 8 hours per day or 12,000 fc hours per year, including daylight.

At the same time, use a filter to reduce ultraviolet (UV). If the fixture is designed for a lens, get a glass or acrylic UV filter. (Do not install a lens in a fixture that is not supposed to have one. It could get too hot.) If the fixture is open, get an MR-16 or -11 source that disposes of ultraviolet with its own dichroic lens. The lens rejects ultraviolet wave lengths, allowing all others through.

One UV protected MR-16 comes in 20 or 50 watts and has 10-, 24-, or 36-degree beams. The 20-watt MR-11 has 10- or a 30-degree beams. Chose the width that fits!

Another UV protected MR-16 is infrared-enhanced with color stablility and greater intensity. An infrared coating redirects the heat inward, producing more light. They are 40 percent more efficient. The spot source can produce 2,300 more candlepower. The 20- and 50-watt beam widths are 10, 24, and 38 degrees.

Second, what sources can provide light? Line-voltage and low-voltage can. If line-voltage, the choices are PAR or T. If low-voltage, the choices are PAR, AR, or MR. (AR can be a small reflector or a large PAR-36 shape.)

For the brightest light, use halogen sources. They give a brighter, whiter light than non-halogen and last longer.

Open MR sources are sensitive to having the filament capsule touched with bare hands—oil from hands could cause the capsule to shatter when lighted. Thus, always specify glass-covered, not open, MR sources and avoid the hazard of having your clients showered with glass slivers!

Low-voltage sources were designed for accenting. They have an accurately controlled beam, a selection of beam widths, and a greater intensity of light for the watts consumed. Not a bad package!

A low-voltage fixture can be recessed, surface-mounted, or clipped into a track. The fixture must be adjustable in order to aim the light at the painting. Recessed fixtures normally are adjusted internally. Surface- or track-mounted fixtures are adjusted externally.

Low-voltage recessed fixtures can fit where standard sizes cannot. They are inconspicuous. Some have built-in transformers. Surface-mounted fixtures do not, and they can be used with simple wiring. Thus, they are easy to use when relighting an existing structure.

At night, an oil painting fades from view.

Turning on lamps does not help.

Surface-mounted fixtures accept only regular reflector bulbs. Surface-mounted and track fixtures adjust the most vertically. However, some are more adjustable than others.

A track fixture can hold a heat filter to fur-ther reduce radiant heat. If you do not mind seeing the hardware on the ceiling, track equipment is a good choice for lighting paintings. If you do mind seeing the hardware, use a recessed low-voltage fixture.

track equipment

Track equipment looks theatrical. It consists of an electrified holder, a low-voltage fixture, a low-voltage bulb, and a connection mechanism for the electricity. A connector mechanism is either an electric junction-box cover or a cord for plugging into a baseboard receptacle. An electric junction-box cover can be at the end or anywhere along the track. The junction-box connection is permanent; the cord-set connection is temporary and less attractive.

Electrified Holder

The electrified holder is either a track for several fixtures or a canopy for one. Electric tracks are available in 2-, 4-, 8-, and 12-ft (0.6-, 1.2-, 2.4-, and 3.6-m) lengths and some can be cut shorter on the job. They are available in single to quadruple circuits. In residences, single circuits are used primarily; in commercial spaces, double, triple, and quadruple circuits are used when many items are accentuated at one time, or when changes of accenting are needed by the flick of a switch. Choose the number of circuits needed depending on how the track is to be used.

- If you want to light just the painting and are likely to move it on that wall, choose a single-circuit track and one fixture.
- On the other hand, if you want to light just the

painting and never move it, choose a canopy and a track fixture. A canopy accepts a single track fixture and permits the fixture to adjust up, down and around, but not slide side to side. It creates a surface-mounted low-voltage fixture out of track components. The canopy is round, about 5 in. (13 cm) in diameter. It is installed directly into a ceiling junction box and wired to a wall switch. A canopy accepts any low-voltage fixture.

- If you want to light the painting and the wall at the same time, choose a single-circuit track and several fixtures.

- If you want to light the painting sometimes and the wall at other times, choose a double-circuit track, several track fixtures, and two switches.

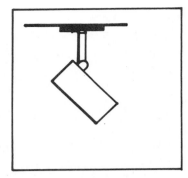

Canopy track light.

A canopy cover can substitute for an electric track.

Low-Voltage Fixtures

All low-voltage fixtures need transformers to reduce the regular line voltage. The transformer is either built into the fixture or installed separately from the fixture—called a remote transformer. Any transformer requires approximately 20 percent additional electricity to function. Low-voltage track fixtures can contain their own transformers.

Low-voltage track fixtures are made in black, white, brass, aluminum, or bronze. Choose a color to match the ceiling and reduce the visual impact, or to contrast and create a visual image. Either way, purchase all fixtures from the same manufacturer. Components are not interchangeable.

Low-voltage fixtures are also made in recessed or surface-mounted styles. Some local codes require that the low-voltage transformer be accessible. Consequently, some fixtures are made to have the transformers within the housing itself. Low-voltage fixtures accept either multifaceted reflector (MR) or parabolic-aluminized reflector (PAR) sources. They are engineered to be wall-washers or adjustable. Wall-washer apertures are black baffle or specular cones. Adjustable apertures are baffles, drop-down mirror, eyeball, or gimbal rings.

Both circuits "on" in a two-circuit system.

Adjustable recessed fixtures have elegant capabilities for lighting art with minimal source recognition. The larger the light source chosen, the larger the aperture required for the fixture. Well-engineered fixtures do not waste light when tilted toward the art. Determine the tilt required. Not all fixtures tilt sufficiently.

Low-Voltage Sources

A low-voltage fixture requires a low-voltage source. Low-voltage halogen PAR, AR, and MR produce precise beams of light. PAR beams vary from 5 to 30 degrees in 35 or 50 watts. AR beams vary from 6 to 32 degrees in 15 to 50 watts.

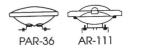

PAR-36   AR-111   MR-16   AR-48

Reprinted by permission of Osram Sylvania.

MR beams vary from 7 degrees for a very narrow spot (VNSP) to 55 degrees for wide flood (WFL) in 20 to 73 watts.

MR-16 BEAM WIDTHS

very narrow spot (VNSP)	7 to 9°
narrow spot (NSP)	9 to 15°
spot (SP)	20°
narrow flood (NFL)	24 to 30°
flood (FL)	38 to 42°
wide flood (WFL)	55°

MR's have either delicate pin bases or more user-friendly footed bases that twist and lock. MR's are either open or covered in glass. MR fixtures are either open or have a lens. If fixtures are open, definitely specify a covered MR-16 to eliminate the bulb shattering when turned on if it was handled by bare hands. Glass-covered MR's are the safest for the general public.

MR's are either standard or color-stable sources. Color-stable sources are also offered with infrared technology (a reflecting film to trap wasted invisible infrared), increasing light output.

Some MR-16's replicate daylight colors, making a difference in the appearance of the lighted art. They are 47K. Others have color temperatures of 3,050°K and 3,200°K. Life of MR's varies from 3,000 to 5,000 hours. Choices; choices!

Compare performances. One comparison could be the amount of illumination created by a standard and an infrared MR-16 source, both with a life of slightly over 3,000 hours. (AR has a 2,000-hour life.) Compare center beam candlepower (CBCP).

COMPARE THE PERFORMANCE
OF STABLE-COLOR MR-16 @ 50 WATTS

SOURCE TYPE	BEAM	CBCP
infrared	spot	16,000
non-infrared	spot	11,000
infrared	narrow flood	6,000
non-infrared	narrow flood	3,200
infrared	flood	3,000
non-infrared	flood	1,800

The qualities desirable for MR sources are: no color shift throughout source life, no discoloration at edge of beam, no discoloration of reflector, and no strange color hues.

Line-Voltage Sources

If brilliance is not an issue, BR (reflector) sources are possible. However, for the same energy consumption, a halogen PAR will provide more light.

COMPARISON OF PERFORMANCE
BR AND HALOGEN PAR @ 60 WATTS

SOURCE	BEAM	CBCP
BR-30	FL	510
	SP	1,600
PAR-30	FL	3,600
halogen infrared	SP	16,000

Clearly, PAR yields the most light for the watts. Be aware, however, that it has a harder edge of light. BR has a softer edge. This may be a design consideration.

PAR   BR

PAR's are PAR-20, -30, and -38 with short or long necks, fitting different adjustable fixtures—recessed or track. Long-neck PAR-20 and -30 are designed for existing fixtures that were originally made for R sources. Short-neck PAR's can be adapted to deep fixtures with a screw-based extender.

appearance of lighting

Two sets of factors affect how the lighting appears from any low-voltage fixture. One set is the size of the lighted area, and the other is the intensity of the light.

Size of the Lighted Area

The size of the lighted area depends on the beam width, the fixture's distance from the painting and aiming angle, and the amount of illumination in the room. First, beams get wider as they get farther away from the light source. For example, a PAR narrow spot can be 1 ft (0.3 m) wide at 6 ft (1.8 m) away, and 2 ft (0.6 m) wide at 12 ft (3.6 m) away. Therefore, the closer to the painting, the smaller the lighted area is; the farther away, the larger.

The beam width originates at the light bulb. Both fixture and light bulb manufacturers publish catalogs with lighting performance data describing beam widths of bulbs at various angles.

Second, the fixture's distance from the painting and the aiming angle determine the size of the beam at the painting. These factors are interrelated. The aiming angle is measured in degrees from something. The aiming angle for a surface is 0° when the light comes perpendicular to that surface. Therefore, the aiming angle for a vertical surface (a wall-hung object or a display mannequin) would be 0° when aimed perpendicular to or directly to that surface. The aiming angle for a horizontal surface (the floor or a tabletop) would be 0° when aimed straight down. However, this definition is not consistent among manufacturers. Some consider 0° always as straight down, and others consider 0° as perpendicular to the object being lighted. Make sure which system is being used when consulting the lighting performance data in catalogs. Otherwise, what is 30° angle in one system is 60° in another; 45° is the same in both.

In this book, the aiming angle is considered 0° when it is perpendicular to the object being lighted. Therefore, the greater the angle, the closer the fixture will be to the painting and the smaller the distance from the wall. The smaller the angle, the farther the fixture from the painting and the greater the distance to the wall.

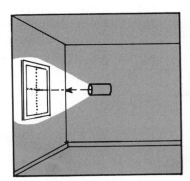

The closer to the painting, the smaller the lighted area.

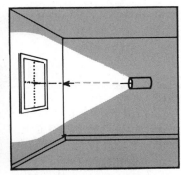

The farther away, the larger.

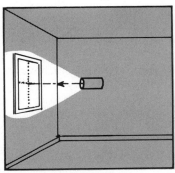

0 degree aiming angle.

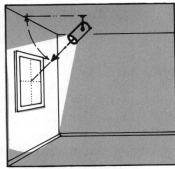

45 degree aiming angle.

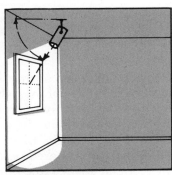

60 degree aiming angle.

Third, the size and amount of illumination in the room affect the area of light seen on the painting. If the wall is brightly lighted, the area of light on a painting will not be definable; it will be washed out by the other light. If the wall is dim, the area of light will be more distinguishable. Decide how dramatic the area of light is to be.

Rule of Thumb for Attracting Attention

In order to attract attention, the light on an object must be three times more intense than its surroundings; ten times the intensity will definitely be noticed.

Intensity of the Lighted Area

At any one aiming angle, the size of the area lighted is approximately the same for both the 25- and the 50-watt PAR bulbs, but the intensity of light is not. The difference in intensity is from two to five times as much. Of course, the intensity is also affected by the distance of the fixture from the painting. As the distance increases, the intensity decreases, because the further light travels, the more it diminishes. This phenomenon is called the Inverse Square Law. It makes the light from the sun tolerable; otherwise, the sunlight would burn us to cinders.

Likewise, the intensity seen is affected by the amount of light in the rest of the room and the color of the painting. The brighter the light

GUIDE TO LIGHTING EFFECTS

	EFFECTS	
FACTORS	Intensity of Light	Size Lighted
Very narrow spot bulb	brightest	smallest
Narrow spot bulb	to	to
Wide flood		
Very wide flood	least bright	largest
Close to painting	brighter	smaller
Far from painting	least bright	larger
Light direct on painting	brighter	smaller
Light at an angle	less bright	larger
Bright room lights	makes less visible	
Dim room lights	makes more visible	
Dark painting	makes less visible	
Pastel painting	makes more visible	

in the room, the less intense the light will appear on the painting. Conversely, the darker the room, the more intense the painting will appear. Dark paintings absorb more light than pastel paintings, thereby requiring more light to be illuminated. In spite of all the factors that affect the amount of light, just 50 footcandles will illuminate a painting sufficiently, in many cases. Yours, however, may be different.

The intensity of low-voltage light can be dimmed with an electronic low-voltage dimmer or a variable autotransformer.

lighting design

Most artwork should have the lighting aimed at a 45° to a 60° angle from perpendicular for the best effect. Glossy artwork (using shiny oils, acrylics, or glass frames) might require a steeper angle (60° to 70°) to avoid glare from the light. However, light at such an angle creates deep shadows and might emphasize too much texture in artwork. Check first to see if the emphasis is pleasing.

Lighting Just the Painting

When lighting just the painting, it is possible to superimpose the shape of the lighted area within the picture frame only, thereby making the painting appear to radiate light. It can be done with a low-voltage PAR bulb, because the shape of the lighted area is square to elliptical. A framing projector can do the same thing but requires

more watts to do it. (See Chapter 8.) To illuminate just the painting with any low-voltage bulb, determine the exact location for the fixture, because the distance determines the size (and intensity) of the lighted area. The best method to use is calculating carefully beforehand and making minor adjustments at installation time. The next best method is to see for yourself.

See for Yourself: What Is the Size of the Lighted Area?

1. If your fixture accepts a PAR bulb, have an electrician equip it with a 50-watt wide flood.
2. Wire the fixture by connecting it to an electric cord and plug with alligator clips, carefully wrapping the clips with electrician's tape and not touching them.
3. Hold the fixture at the ceiling where you could install it.

4. Move it forward or back to get the desired area of light.

5. If the area is too small with the wide flood, try the very wide flood. If it is too large, try the narrow spot or very narrow spot.

6. Make sure that glare is not going to be reflected from the artwork.

7. Mark the place on the ceiling where the fixture must be installed. (If a sofa or chair is in front of the painting, make sure that glare would not get into the eyes of a seated person. If so, install the fixture closer to the wall—at a steeper angle—and use a bulb with a wider beam.)

Light the Painting and the Wall

To illuminate your painting and the wall around it, locate the fixture so that the light comes from a 60° angle. For example, with a painting centered at 5 ft 3 in. (1.6 m) above the floor, install a track 1 ft 8 in. (0.51 m) away from the wall in 8-ft (2.4-m) ceiling heights; in 9-ft (2.7-m) ceilings, install the track 2 ft 3 in. (0.69 m) away; in 10-ft (3-m) ceilings, install 2 ft 9 in. (0.84 m) away. After installation, minor adjustments of up, down, left, or right can be made by moving the fixture itself. Track equipment is adjustable.

Sample Electric Cost

A low-voltage fixture consumes significantly less electricity than a regular line voltage fixture. For instance, using a 50-watt low-voltage bulb for 4 hours per night at 10¢ per kilowatt-hour, the electricity charge (including 20 percent more electricity for the transformer) is 72¢ for a month. By contrast, a standard voltage bulb, to do the same job, would need to be at least 75 watts and usually 150. As a result, a 150-watt bulb used for the same time period and at the same rate would cost $1.80. In addition, it would generate more heat, which might be uncomfortable during the hot weather or might damage the painting over a long period of time. Would you rather pay 72¢ or $1.80 per month to accomplish the same thing? A low-voltage fixture allows you to light up your painting and not your utility bill.

Dramatically light your painting within the frame only.

Light, less dramatically, both the wall and the painting.

putting your collection in the light

11

Collections are accumulated and displayed. But if they are not lighted, they do not get the attention they deserve. Whatever objects you collect—trophies, antiques, prints, sculpture, family photographs, or others—light can enhance their special characteristics.

what do you collect?

- Metal objects; light makes them shiny and glimmering.
- Glass; light is essential to give sparkle and brilliance.
- Ceramics; light reveals shape and subtle colors.
- Cloth, leather, or wood products; light points out texture and grain.
- Sculpture; light gives form, highlight, and shadow.
- Wall-hung art (photographs, prints, bas relief); light reveals colors and adds richness and authority otherwise not present without light.

In addition to complimenting a collection, the room itself benefits from light. It bounces into the room and softly balances the light from other sources. Light from a lighted collection can brighten difficult places, such as dark corners and dim sidewalls.

collections displayed on furniture

If furniture displays collections, purchase it already lighted or have it custom lighted. Already lighted furniture usually has incandescent light sources. However, consider replacing them with more energy-efficient fluorescent sources—new small screw-base bulbs, or tubes on a track or in miniature fluorescent holders. Fluorescent sources last longer than do incandescent sources and produce less heat that could damage collectibles.

Ready-made furniture such as china cabinets, breakfronts, secretaries, bookshelves,

Light reveals shape and subtle colors of ceramic collections.

Light a breakfront displaying Spodeware for
$1.24 a month at $.10 per kilowatt-hour.

étagères, and headboards are available with built-in lighting. Lighted furniture is excellent for both owners and renters. Furniture stores and interior designers feature them. However, ready-made furniture usually requires incandescent bulbs—sometimes tubulars (like the ones in showcases of department stores) and at other times reflectors (like a spotlight). Tubulars accent objects best from the front. Reflectors accent best from the top (downlighting) or from the bottom (uplighting).

The color of the incandescent light enhances the look of the wood in the cabinetry but detracts from the colors of blue and green objects. More important, heat is produced, which could fade colors or damage objects, particularly paper, leather, or other nondurable goods.

As energy gets scarcer, furniture manufacturers will offer cabinetry with more efficient light sources. In the meantime, if cabinets are chosen with tubular incandescent lights, seek a reflectorized bulb, which spreads the light. They are hard to find but worth the effort. If it is unobtainable, back up a tubular bulb with metal, spreading the light and protecting the cabinetry from the heat. Likewise, if the cabinet has a reflector bulb (flood or spot), choose glass shelves to allow the light to go all the way through; otherwise the watts are wasted on just one shelf.

Whereas a 20-watt incandescent would illuminate the center of a shelf with lots of watts and heat, a linear fluorescent could spread the light farther with less watts and heat. A T-2 fluorescent source could do the job. T-2 is $\frac{1}{4}$ in. (6 mm) with 11 or 13 watts, producing 680 or 860 lumens. The fixture is $\frac{3}{4}$ inch (19 mm) deep, mountable on wood or glass shelves, hard-wired or plugged in. It has a reflector for an asymmetric distribution (120°). Lengths are 20, 24, 37, and 47 inches (51, 61, 94, and 119 cm) long, producing 680 to 1,720 lumens. Hard-wired T-2 sources can be installed with a remote ballast by an electrician.

If a higher level of illumination is required—dark cabinets with dark objects and high ambient lighting, use T-5 ($\frac{5}{8}$ in. or 16 mm) sources, producing up to 3650 lumens. T-5 fixtures can be shelf-mounted tracks that accept both fluorescent and incandescent sources, for highlighting special objects. Or they can be asymmetric fixtures. Either way, purchase a length that fits the shelf from side to side.

- For backlighting, tuck them neatly behind or below the shelves.
- To accent, place them on the left or right, inside the cabinet door.
- To uplight, secure below a frosted glass shelf, to hurl their light up.
- For frontlighting, slip them beneath the shelf above, at the front.

Whatever the position, hide the fixture behind a structural portion of the cabinet or a piece of wood trim added on.

Installation and Effects

Different materials are enhanced by light from different locations. Metal objects—guns, trophies, sculpture—respond better to light from the sides to accentuate the surface of the metal and enhance the shine. Glass objects—bottles, boxes, ornaments—respond better to light from below or above. The edges of the glass will take on the light and glow brilliantly. Glass objects on lighted frosted glass glow as though they were bursting with luminescence. Ceramics, cloth, and leather respond better to light from the front, above or on the sides, revealing the shape, color, and texture. If the color is not important, backlighting reveals the silhouette and calls attention to the shape of the object.

Sample Electric Cost
The light for a breakfront with three shelves consists of six fluorescent ready-made fixtures of 15 watts each, totaling 103 watts, including 15 percent more electricity for the ballast (the built-in regulator). The light for 4 hours per night at 10¢ per kilowatt-hour costs $1.24 per month.

---

free-standing collections

Free-standing collections are three-dimensional objects that stand alone, such as sculpture, large antique vases, and oriental screens. Free-standing objects have unique forms, solid or open, which require highlight and shadows to reveal them. Light creates highlights as well as shadows. Without light, objects can appear flat and uninteresting. Only direct light gives birth to highlights and shadows—and it must be powerful. It can also be energy-efficient if it comes

Freestanding objects demand light.

Installation

The light from the front and sides can originate at a ceiling with a low-voltage fixture. The light from behind can also; or else it can be an uplight with a standard-voltage fixture, provided it does not glare in anyone's eyes.

A ceiling fixture can be recessed, surface-mounted, or on a track or canopy. The track or canopy can be installed either permanently into a ceiling junction box or temporarily hooked up by a cord and plug to a baseboard. A surface-mounted or recessed fixture, on the other hand, must be permanent. Owners can permanently or temporarily install fixtures or tracks, and renters can install tracks temporarily. A floor uplight can sit on a weighted base and can be hooked up by a cord and plug. It is a temporary installation for both owners and renters.

Fixtures

Low-voltage fixtures are adjustable. A transformer (an apparatus to change the voltage of electricity) is built into the fixture and reduces the standard current (120 volts) to 12 or 5.5 volts. The fixtures are manufactured in compact shapes, and they utilize low-voltage bulbs. They make it possible to capture the best highlights and shadows, which is very dramatic and artful.

Sources

The best sources are halogen, low-voltage, either MR, AR, or PAR. MR puts out very narrow spot up to wide flood beams. MR's are slightly over 30K or 47K—daylight-like in color. Their brightness ranges from 560 center beam candlepower (CBCP) to 16,000, in 20 to 73 watts. Some have infrared technology to increase light output. Some control color shift and decrease ultraviolet light.

from low-voltage fixtures and bulbs. With low-voltage fixtures, heat is also low and the bulbs last longer.

Each free-standing collectible must be tested with light to determine the best way to create the highlights and shadows. One light is unlikely to make an object appear dynamic; two or three are frequently needed.

See for Yourself: What Will Give the Best Effect?
1. Use a flashlight, an auto trouble light, and any portable lamps you have.
2. Shine the brightest on the front from above, one from the side, and another from behind.
3. Move them until you find the positions that create the greatest visual impact.

Not all of the light should be equal. Some lights should be brighter, some softer. Consequently, after installing the fixtures, try several wattages, using the least first, and working up to a higher wattage if necessary.

Sample Electric Cost
A sculpture highlighted by two multifaceted reflector 50-watt low-voltage bulbs for 4 hours each night costs $1.44 per month, at 10¢ per kilowatt-hour, including 20 percent wattage for the transformer. Low-voltage fixtures cost one-third more to purchase than do regular voltage fixtures, but they consume two-thirds less electricity over the long haul.

---

collections hung on the wall

Collections that hang on the wall include photographs, prints, paintings, tapestries, macramé, and bas-relief. How to light them depends upon how serious the collector is, where the collection hangs, how often it is changed, and whether the collector is an owner or a renter. A serious collector will be willing to spend more to enhance his or her collection. A casual collector will want enhancement too, but at a smaller cost. A permanent collection on one wall can be lighted with wall-to-wall permanent lighting. A changeable collection requires changeable lighting. Owners are able to build in lighting. Renters must use removable lighting.

Owners with Serious Collections on One Wall

Owners with a serious collection hung on one wall will want to illuminate the whole wall with energy-efficient fluorescent wall-washing. It will flatter the collection, the room, and the occupants, particularly when equipped with deluxe or prime-color fluorescent tubes. (See built-in fluorescent cornice light in Chapter 5.)

The collection can be changed without disadvantaging the lighting effect. If the collection happens to be framed with glass, fluorescent light illuminates it glare-free. If the collection is removed completely, the lighted wall will continue to flatter.

Owners with Serious Collections on Several Walls

Owners with a serious collection spread out on many walls should check Chapter 10 for various methods of lighting one piece of wall-hung art. Whatever method is chosen, select the best items in the collection to be illuminated and also decide which rooms could benefit from the extra light. In most cases, extra light helps semi-public rooms: in residences—living rooms, family rooms, dining rooms, and hallways; in businesses—waiting rooms and executive offices.

Some owners may want to invest in a low-voltage track canopy and track fixture. The track fixture needs to be adjustable but does not need to slide. The track fixture could be detached from the canopy and put away if the collectible is removed. Purchase a low-voltage track fixture and a track canopy for each collectible to be lighted. Equip the fixture with a low-voltage bulb in the beam width that matches the size of the wall hanging. Try the 20-watt PAR flood first if the wall hanging is not large. Otherwise, try the 50-watt very wide PAR flood. No rule of thumb is available for lighting collections, because so many variables need to be considered—color of the object, distance from the object, and the total light in the room, to mention a few. (See Chapter 10.)

INSTALLATION AND EFFECTS

The size and brightness of the area lighted depends first on the type of bulb chosen and second how far away it is. The closer the bulb, the smaller and brighter the lighted area is. The further away, the bigger and softer it is. The bulb will produce effects ranging from a bright rectangle of light to a soft, pool-shaped glow.

An electrician should pretest the track canopy before deciding where to install it. Typically, it should be installed about 3 ft (0.9 m) out from the wall and centered on the collectible. Make final improvements after the fixture is clipped in to the canopy. Test the three wattages and three bulb types available. One will satisfy.

If the collection is framed with any kind of glass, avoid creating a glare. Glare obscures—especially on photographs where the details are small. Glare can be avoided by installing the fixture as close to the wall as possible and grazing the picture with light.

Wall-hung objects can be lighted by a cornice light.

Owners as Casual Collectors

Owners, with a casual collection, either changing or static, should use a fluorescent wall bracket. Mount it high at the top of the wall or just above the wall object. Brackets effectively light wall objects up to 8 ft (2.4 m) wide or a group of smaller objects. The bracket must be larger than the object, but not much larger. Brackets must conform to the standard increments dictated by the sizes of fluorescent tubes: 4, 6, or 8 ft (1.2, 1.8, 2.4 m). Ready-made wall brackets are available, in 4-ft (1.2-m) and 8-ft (2.4-m) sizes. They are either box-shaped or tube-shaped. They are available in many colors. They mount flat to the wall, or hang by a stem from the wall or ceiling. On long walls they can hang in tandem, stretching in modules divisible by their length.

Likewise, fluorescent wall brackets can be custom-made. They are usually box-shaped and built like a dropped soffit, but narrower, using only one bulb. Therefore, a bracket is built approximately 6 in. (15 cm) wide. (Follow the directions in the dropped soffit section in Chapter 9.) Custom-made brackets can be positioned at the top of the wall or directly above the object. Because the 4-ft (1.2-m) fluorescent bulb is the easiest to obtain and the quietest with a good-quality ballast, build your bracket to use this size, if possible. If a 4-ft (1.2-m) size is too small and two 4-ft (1.2-m) fixtures are too big, try 3- or 6-ft (0.9- or 1.8-m) sizes.

INSTALLATION

Center the bracket over the object. Avoid a cord and plug connection, if possible. Install the bracket through a junction box in the wall or ceiling, provided that electric wires are accessible. Owners can specify a junction box when developing the building plans, or when relighting they can have an electrician gain access in an attic or pull wires from elsewhere.

Renters with Serious Collections on One Wall

Renters with a serious collection on one wall can light it with a track and several low-voltage fixtures. The track can be electrified by a cord and plug and can be mounted on the ceiling for the duration of the lease. At moving time, it can be moved to illuminate another wall in another location.

FIXTURES AND BULBS

Select only low-voltage bulbs in the sizes that produce the amount of light you want. It is costly to dim low voltage. It must be done through the primary electrical source to the transformer and requires a special low-voltage dimmer.

In making the bulb selection, decide if an even wall-to-wall wash of light or variable pools of light is wanted. For even light, space the fixtures equally and equip them all with the same size bulbs. For variable light, space the fixtures as dictated by the position of the objects on the wall. Equip the fixtures with either the same bulbs for equally bright pools of light or different bulbs for brighter and softer pools of light. A mixture of spots, floods, or very wide floods produces different amounts of light for a variably lighted wall. (For bulb sizes and brightness amounts, see Chapter 10.)

INSTALLATION AND EFFECTS

Check local electrical codes to determine if a cord and plug are permissible. If so, position the track about 3 ft (0.9 m) from the wall, running the cord along all the corners—ceiling, wall, and baseboard. This installation makes a temporary connection look permanent; it is the least obtrusive way. Purchase enough cord to reach the baseboard receptacle by way of taking the cord to the closest wall, moving along the ceiling-wall corner, to the wall-wall corner, then down to the baseboard and over to the receptacle. The added expense and effort are well worth the added results. Finally, adjust the fixtures to illuminate either evenly or variably. Even lighting creates a broad wash of light at the intensity deliverable by the wattage chosen. Variable lighting emphasizes lighted objects, not the wall. Either type of lighting is appropriate; both are spectacular.

Renters with Collections on Several Walls

Renters with either serious or casual collections on several walls should use portable lamps—floor lamps, table spots, picture lights, or possibly pin-ups.

FLOOR LAMPS

Floor lamps sit on the floor, reaching out to highlight a nearby wall object. Some floor lamps have multiple arms to spot several objects at once. Some have multiple lights but not arms, to spot not quite as far. Some are slim shafts of fluorescent light. If positioned in pairs, they can illuminate wall objects from two sides.

TABLE SPOTS

Tabletop spots sit on top of furniture or clip on a shelf, aiming at a wall object. Tabletop spots are scarcer than floor lamps and are mostly slick contemporary shapes; some blend with traditional furnishings. They accept standard or low-voltage bulbs. The type needed is determined by the distance away from the object and the brightness desired. The longer the distance, the higher the wattage required. To be energy-effective, place the spot lamp close to the object on shelves or tabletops next to the wall. In most situations, a spot lamp with a 50-watt standard voltage reflector bulb does the job.

PICTURE LIGHTS

Picture lights illuminate pictures by hanging on the wall or on the frame. They use tubular incandescent bulbs. The tubular R-type bulb is the best. It has a silver reflector strip that allows all the light to be directed to the picture. Picture lights are available in a variety of sizes, most often 10, 18, and 28 in. (25, 46, 71 cm) long. Likewise, pictures can be lighted on an easel away from the wall. Either way, the power of picture lights is limited. They throw noticeable light down only 3 ft (0.9 m). Therefore, plan to light a picture that is no more than 3 ft (.9 m) high. Further, choose a light that is at least two-thirds as wide as the picture. For example, use a 10-in. (25-cm) picture light for a 16- by 36-in. (41- by 91-cm) picture. When deciding to buy, remember that within the limits of its candlepower (the ability to produce light) the farther out the picture light extends from the picture, the farther it throws the light down.

TRACK PICTURE LIGHTS

Some track lighting fixtures can light pictures.

Tabletop spots.

Use one or two 25-watt reflector incandescent bulbs. The track can be mounted above the picture on the ceiling or the wall. Adjust the fixture and aim the bulbs to provide the best light.

Sample Electric Cost

For 4 hours a night at 10¢ per kilowatt-hour, these two 25-watt bulbs consume 60¢ worth of electricity per month to illuminate one picture in a collection.

PIN-UPS

A pin-up (a cord-and-plug connected, wall-hung fixture) can be a picture light in some cases. A pin-up is not always aesthetically successful as a picture light, because it is a large and obvious gadget. However, pretest it in your space and maybe it can be arranged to be artistic and harmonious.

At most, picture lights throw light down only three feet.

protecting collections from fading

Collections are costly, some more than others. Owners of collections either want them brightly lighted so that they may enjoy them in their lifetime without worrying about the future value, or owners want their collections lighted but protected to be passed along as heirlooms. Both positions are defensible. Whichever your client chooses can be accomplished.

If the goal is to light but also to save, guidelines have been developed by the lighting engineers at the Smithsonian Institution. Their guidelines cover a wide range of materials found in collections, because surprisingly enough, watercolor and prints are not the only collectibles subject to depreciation due to light. Even glass can be damaged (old glass more than new glass). On the other hand, new paper (pulp paper) is easier to damage than old rag paper.

The destructive aspects of light affect many materials. The destructive aspects are the ultraviolet and the infrared parts of the spectrum. Ultraviolet (UV) fades and deteriorates. Infrared dries out. Consequently, light-source type and amount of light need to be chosen for the least possible damage.

LIGHT SOURCES

SOURCE	APPROXIMATE % OF ULTRAVIOLET LIGHT
north sky daylight	over 20
sunlight	10–20
all daylight	10–20
mercury, coated bulb	10–20
metal-halide	10–20
fluorescent, deluxe	12
fluorescent, cool-white	8
fluorescent, warm-white	4
tungsten halogen	8
fluorescent, rare earth/triphosphors	4–8
low-voltage incandescent	4–8
standard incandescent	4
sodium	under ½

Reprinted by permission of the Smithsonian Institution.

In many residential and in some commercial spaces, daylight is a light source. It must be counted as part of the total light affecting collectibles. Light can destroy and daylight is actually more destructive than electric light. One of my clients wanted to light her extensive wall-hung collection of delicate artwork, but it was already fading from being exposed to north, south, east, and west daylight. (She had a living room with windows on all four walls letting in Florida sunshine.) Consequently, I specified only narrowly confined incandescent sources and rare-earth fluorescent to light the space.

Amount of Light

All materials depreciate with light, some more than others.

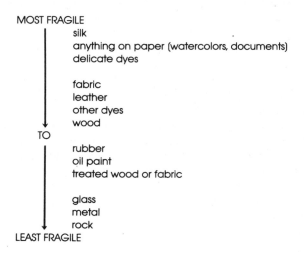

MOST FRAGILE
- silk
- anything on paper (watercolors, documents)
- delicate dyes

- fabric
- leather
- other dyes
- wood

TO

- rubber
- oil paint
- treated wood or fabric

- glass
- metal
- rock

LEAST FRAGILE

Overall, how much the material dries out, deteriorates, or fades relates to how long the collectible is exposed, the intensity and spectral characteristics of the light, the change in humidity and the chemistry of the material. Ed Robinson of the Smithsonian says that the chemistry of some pulp paper is so internally destructive that even in the dark some paper destroys itself. With light, the process is faster. (Consequently, in 1986, the White House started using rag paper for all the President's paper—scratch, note or letter—since it could be historically valuable.)

In general, the most fragile collectibles should not be exposed to more than 12,000 footcandle hours per year; that is, about 5 footcandles times 8 hours per day for 300 days per year. This rule assumes that all ultraviolet is blocked. It can be blocked successfully by filters. Filters can be glass or acrylic, or the light source itself can be tinted to filter UV. Determine how much time the object will be exposed to light, then determine the footcandles allowable. Err on the side of less-light rather than more-light, if preservation is the objective.

read in bed
but
do not disturb
12

Commands of "turn out that light" and retorts of "I only have one more page" echo in many bedrooms, because some people enjoy reading in bed and others want the lights turned off. This conflict is a lighting problem that most of my lighting clients ask me to solve.

Light for bedtime reading usually comes from lamps on bedside tables. But bedside lamps do not illuminate reading material well enough and definitely do not confine the light enough to permit one person to sleep while the other reads. Further, bedside lamps are not usu-ally adjustable and cannot be moved to permit the most comfortable reading position. Likewise, in new construction, room sizes are often too small to accommodate bedside furniture with lamps. This is no great loss. Wall fixtures are actually better. Adjustable wall fixtures permit one person to read while the other sleeps, if they have opaque shades (shades that do not transmit light). Use a single, a double, or two single fixtures. Install them at the head of the bed so the light can be directed away from the sleeping partner.

installation

Wall fixtures are either wired directly through the wall or plugged into a baseboard receptacle. The through-the-wall method requires an electrical junction box. This method is called either wall mount or outlet-box mount. (Manufacturers' terminology differs.) The plugged-into-the-wall method requires a cord and plug, and it is called either pin-up or cord and plug. When ordering the fixtures, specify which type you want. Owners can take advantage of outlet-box mount; renters must resort to the cord and plug. An outlet-box mount should be incorporated when building and can be incorporated when relighting. When building, owners should specify an appropriately placed junction box before the building process begins. When relighting, have an electrician pull wires from a baseboard receptacle to electrify a junction box. In most cases, an electrician is required to install an outlet-box mount fixture. Nonetheless, it is the slickest method, because wires are not visible.

If outlet-box mounting is not possible, purchase a fixture with a cord and plug. The cord usually hangs below the fixture, unless you are clever about hiding it or unless you choose a fixture that has a tube to hide part of the cord—

One person can read and one can fall asleep
with a wall-mounted fixture.

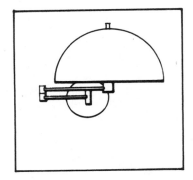

Wall-mounted fixture.

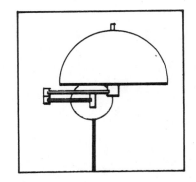

Pin-up fixture.

appropriate on some styles, inappropriate on others. No electrician is needed for this installation, and renters find this method the best.

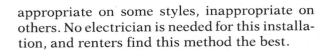

Rule of Thumb for Mounting a Reading Light
Mount the wall fixture 30 in. (76 cm) above the mattress.

Modify this rule if necessary to make sure the fixture is within easy reach from a semireclining position, in order to be able to adjust it or turn it on and off. In addition, the rule may need to be modified according to the headboard height. Center the fixture above the bed, and point the light to the reader's side. Finally, for the greatest convenience, locate the on-off switch either on the fixture or on the wall plate behind the fixture.

fixtures

Choose a fixture with a shade that does not allow light to escape—usually metal. However, metal absorbs heat from the bulb. Some fixtures are designed to dissipate part of that heat through perforated shades, letting light escape also. To confine the light, choose only solid shades or high-quality fixtures that dissipate the heat and not the light.

Select an adjustable fixture—with an adjustable shade, an adjustable arm, or both. An adjustable shade swivels and allows the light to be directed exactly where needed, but be careful to choose one that does not droop. On the other hand, an adjustable arm, hinged or folding, allows the fixture to reach out farther. Some arms reach as far as 34 in. (86 cm).

Choose a fixture style that is in concert with the room. Often, manufacturers make the same style in both outlet-box-mount and cord-and-plug types. Also, they sometimes make a single or a double version in the same style.

In addition, two track lights can be used for reading lights. Track fixtures can be mounted together on an electric track or singly on a track canopy. Both are adjustable. Track fixtures are movable left to right; canopy fixtures are not. Like other wall reading lights, the switch needs to be on or near the fixture.

Tracks are available in 2- or 4-ft (.6- or 1.2-m) lengths and can be cut to match the width of the bed. Even so, electrical tracks may present a hazard. Although the electrical wiring is well recessed, a paper clip or other small metal object can be poked into an open track and instantly conduct the electricity to the person holding it. Therefore, track lights are not a good choice for areas where children and others not thoroughly responsible for their actions are present.

The electrical connection for a track is made through a junction box or by a cord and plug. Most track canopies require a junction box, but some can be ordered with a cord and plug.

light sources

Many fixtures for reading lights are still designed for incandescent sources in spite of the development of more energy-efficient, cooler, and flicker-free fluorescent. Such sources are screw-based, electronic-ballast, compact fluorescent in 9, 11, and 15 watts. They substitute for 25, 40, or 60 incandescent watts. They last 10,000 hours.

Other compact fluorescent can be pin-based with a screw-based, magnetic ballast adapter. The adapter does not get thrown away when the source burns out. It accepts a pin-based replacement. This saves money. They consume 5, 7, 9, or 13 watts, substituting for 25-, 40-, or 60-watt incandescent. The magnetic ballast is less expensive, but tends to flicker when turned on and can be noisy when used. They last 10,000 hours also.

Most of these small size compact-fluorescent sources are 27K, similar in color to incandescent. Some are available in higher Kelvin temperatures. One actually resembles an A source in shape and is available in 11- or 15-watt sizes with a 6,000 hour life.

Sample Electric Costs

A bedside reading fixture with a 15W compact fluorescent source for 1 hour per night for a year costs 53¢ at 10¢ per kilowatt-hour. A 60W A-incandescent source would cost $2.11 and would burn out 6 times before the compact fluorescent burned out.

Incandescent fixtures are designed to accept certain shape sources—tubular (T), reflector (R or PAR), and A. Reflector sources direct light out of the fixture, tubulars do not. Choose a halogen PAR for the greatest amount of light. The 45-watt PAR-16 gives out as much light as a nonhalogen 60-watt PAR-16.

Standard A sources, however, need well-designed reflectors in the fixture to capture the light. Most do not have them. If using an A source, consider using the halogen A (BT, TB, or MB). They last three times longer than standard A sources and the color looks whiter.

Whatever source is chosen, caution your clients to have another light source on in the space, particularly if they are reading for more than a few minutes. The eyes need to be able to focus on a distant object when a person automatically looks up. This distance focus permits the eyes to rest. Otherwise, the eyes will fatigue. Fatigue is not welcomed by anyone.

Osram Sylvania.

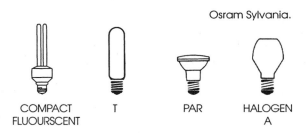

COMPACT FLUOURSCENT   T   PAR   HALOGEN A

hobbies
demand light
13

Hobbies give us a sense of satisfaction and joy. Many are visually demanding. They require sufficient illumination to be accomplished easily. Without light, hobbies become tedious and slow. Adjustable fixtures offer the most options for getting light to a hobby, but do not rely on adjustable fixtures alone; others are needed elsewhere in the space.

needlepoint or other handwork

More light is needed for needlepoint and other handwork than for reading or watching television. More often than not, handwork takes place in the living room or other family relaxation space. Usually, standard light sources—portable lamps and sometimes ceiling fixtures—are the only sources available. But an adjustable lamp that puts out additional light is also needed. If the eyes are not too good or the handwork has little contrast and small details, such as 16-point needlepoint canvas, a 200- to 300-watt standard voltage or 25-watt low-voltage incandescent bulb should be in the lamp. Make sure the lamp has the capacity for such a bulb.

The lamp can be wall-mounted or placed on a tabletop or the floor.

An excellent portable lamp for hobbies with small details is a fluorescent circle lamp with a magnifying glass in the center of the shade. It clamps or sits on a table edge and adjusts to any position. These lamps are excellent gifts for people who enjoy visually demanding hobbies, particularly for those who cannot see as well as they used to. Whichever type of adjustable lamp you choose, always have at least one other light source on in the room to provide background lighting. Otherwise, the eyes can become strained.

woodwork, metalwork, and other workbench crafts

Workbench hobbies may be visually difficult or may be simple. Difficult ones require more illumination. If the contrast of materials used in the hobby is great, the details are large, the length of time is not too long, and the eyes are good, the amount of illumination can be 20 to 50 footcandles. As the contrast lessens, the details get smaller, the time lengthens, and the eyes worsen, the amount of light required goes up. It can go up to 200 footcandles for metal engraving, gem polishing, and jewelry making.

For both day and night, the lighting should be balanced but not bright. The light from a bright sunny window, for instance, should be reduced to balance the amount at the workbench. The workbench fixture should be out of the way—that is, ceiling-, wall-, or shelf-mounted, or suspended. A fluorescent fixture is the most suitable. Position it so that the tube does not reflect an image on any shiny surface—over the front edge or on the two sides of the bench. If mounted over the front edge, the fixture should be about two-thirds the length of the workbench. For example, a 6-ft (1.8-m) workbench should have a 4-ft (1.2-m) fluorescent fixture. Fixtures with reflectors send out more light than those that do not. Ceiling fixtures require a wall switch; suspended, shelf-mounted, and wall-mounted ones do not. Fluorescent fixtures can hold one, two, or four tubes. Choose according to how much light is needed and how much is absorbed by the interior colors. Normally, two tubes are needed, but four may be required for hobbies that are difficult to see. Further, if matching colors is important to the

Workbench fixture should have a reflector.

hobby, special color-matching tubes are available; otherwise, the common cool-white fluorescent is satisfactory. (See Chapter 20 for color-matching details.)

Owners will want to make their fixture installation permanent. Renters can use several inexpensive fluorescent strips with the convenience of a switch on the cord with a plug.

painting, sketching, or collage

If someone paints, sketches, or does collage, colors are critical and the illumination must reveal the correct colors. The amount of light must be sufficient, and the source must contain all the colors in the spectrum. For this reason, artists paint by daylight at a north window. Daylight contains all the colors in the spectrum. The north daylight is not direct or harsh, yet it is sufficient. Not all hobbyists have a north window, and some must paint at night. Therefore,

lighting fixtures must provide the required illumination.

The amount required is as much as in a well-lighted office. The fixtures must light the space uniformly because the hobbyist often copies something elsewhere in the room, and the objects to be copied must also be well lighted. Therefore, ceiling fixtures are suitable; they can be recessed, surface-mounted, or suspended. In addition, for seeing fine details, a light fixture—

pendant or portable—can be located at the easel or the work surface itself.

Should the light source be incandescent or fluorescent? If the end product of the hobby is in color and you know under which light source it will be seen, use the same source. Otherwise, choose the source that works best for the space.

Fluorescent sources can render colors correctly. The color-matching types are used in printing shops and hospitals, where colors are critical. They are more expensive, but like other fluorescent tubes they last a long time and use little energy. (See Chapter 20.)

sewing

Using a sewing machine varies in visual complexity with each sewing job. Many components vary: thread and stitch size, contrast of thread and fabric, and reflectances of fabrics. Maximum light should be at the sewing-machine needle. For complex jobs, 200 footcandles are needed, and the rest of the tabletop should be illuminated to 20 footcandles.

Additional fixtures for additional light need to be located on a wall, shelf, or ceiling, or they can be free-standing. Their position must not create a shadow from the user's hand.

playing a musical instrument

Musical scores vary from simple with large notes and dark lines, to complex with small notes and many line notations. Casual musicians are not as concerned with speed and accuracy in reading a music score as advanced or professional musicians. Casual musicians with simple scores require normal reading light (approximately 20 to 50 footcandles); professionals with complex scores can require as much light as do drafters (approximately 100 to 200 footcandles).

Renters should be aware that portable lamps that mount on the music rack do not illuminate the full page well and should be used only when no other solution is possible. Floor-mounted portable lamps with several light sources can illuminate the score and the general surroundings better.

Owners should be aware that built-in fixtures should be above the musician's head. If the fixture takes an incandescent bulb, it should be adjustable and strike the music rack between a 30°- to 60°-angle from behind the head or on either side of the music. If the fixture is fluorescent, it must not create glare or reflections on the scores.

other hobbies

Follow the illumination requirements similar to those already described in this chapter. If the detail is large, the illumination requirements are low; if they are small, the requirements are high. If the hobby is visually complex, the requirements are high; if it is simple, the requirements are low. For example, jewelry making—very visually complex—requires more light than pottery making—relatively simple. In addition, if 8 hours a day are spent on a hobby, more light is needed than if half an hour were spent. More light is needed for people over 40 years old. Further, if the room surfaces are dark, more illumination is required than if they are pale. Determine whether the maximum or minimum illumination is needed by the requirements of the hobby. Pretest whenever possible and be willing to add more light for your hobbie's requirements when it is necessary.

light for
older eyes

14

As we grow older, our eyes need more light in order to see. After age 55 the eyes do not see as well, sometimes even in good light. Young people think they can see in any light; they often study in a dim room without turning on the available lamp, which is often right next to them. On the other hand, older people know that light acts as a magnifier—they take the needle to the window to thread it or the phone book to the lamp. Older people (particularly those older than 65) should have colorful, detailed, and correct visual information, which can be provided by good light. Some older people claim that they do not need light and that they must save money. Scrimping on light is a false economy. Injury or depression costs much more.

where is light needed?

Ideally, light is needed everywhere. At a minimum, it is needed in the room with the television, at a mirror, at kitchen counters, at stairs, near bathtubs or showers, and in closets. When the television is on, one light should be near but not on top of the television set and another one elsewhere in the room. Television should never be viewed in the dark; it is too tiring for the eyes. Further, light is needed at the bathroom mirror, positioned along the sides to illuminate the whole face. Most bathroom lights are behind the person standing at the mirror, illuminating the top of the head and leaving the face in ghoulish shadow, which is not flattering to anyone.

At kitchen counters, where sharp tools are used, utensils are identified as dirty or clean, and wholesomeness of food is differentiated, light is needed. Yet most kitchens are lighted from the center of the room. This position puts people between the light and what they are doing, causing deep shadows. This is not very safe for anyone at any age.

Light over tubs or showers permits personal care and helps ease getting in and out. Light in closets helps with searching and identifying. Finally, light is needed at steps and other hazardous places, because a potential outcome—personal injury from falls—is so devastating. Many times, older people intend to save money by turning off the light and walking up the stairs in the dark. However, the cost for an injury is infinitely higher than the pennies saved on an electricity bill. After all, a 100-watt incandescent bulb can burn for 1 hour at 10¢ per kilowatt-hour for 1¢. Older people are penny wise but dollar foolish if they think that keeping themselves in the dark will greatly help their budget problems. More money would be saved by heating water less often, because hot water heated electrically averages about 30 percent of the bill and lighting averages about 5 percent. Without adequate light in living spaces, older people are buying into a duller and drabber existence. Light can establish a cheerful environment and can make objects visible for a safe, pleasant old age.

Because the ability to see details has decreased, older eyes need greater quantities of light—about twice as much as young people. Because sensitivity to glare has increased, older eyes need glare-free direct and reflected light. Glare can disable. Because most older folks wear glasses—almost 100 percent over age 65 do—light should not create false illusions of depth. Reflections from a glossy surface can confuse and create illusion. Because contrast sensitivity has declined, older eyes need a suitable relationship of light nearby to light farther away. Contrasting areas of light can cause difficulties in seeing. Because light draws attention, it should focus and guide older people through architectural changes, especially potentially dangerous ones. Because light fixtures need maintenance, fixtures for older people who maintain their own residences should be ser-

Changing bulbs should not require a ladder.

viceable from the floor. Changing bulbs should not require a ladder.

what kind of light should be supplied?

In general, older folks should use soft, diffused light that minimizes shadows, with at least two light sources in each space. Light sources should be large as possible but not bright, such as big lampshades with low luminescence, large opal diffusers, or the largest fluorescent ceiling fixture altered to use fewer tubes than intended. Whenever fluorescent light sources are used, choose warm-white deluxe or prime color tubes. They are preferred by most older people. Whenever incandescent light sources are used, choose shades for lamps that are white or off-white and allow the light to come through. Colored shades impart color to the light.

Interior spaces should be equal in brightness; a bright room should not lead into a dull room. Light should be positioned to guide and call attention to important facilities in a room. No bare light bulbs should be in view. Shield all bulbs with shades or diffusers, especially when seen from above, such as when you are descend-

ing stairs. Use coated incandescent standard A bulbs, not frosted or clear ones. Frosted and clear bulbs have a hot spot; coated bulbs do not. Use strong directional light—spot or flood bulb, accent or track fixtures—only to illuminate sewing or other small details for hobbies and handwork. Do not use a brightly lighted work area in a dark room; light up the room, too.

Do not place electric cords where they can be tripped over. Use tabletop dimmers so that control is at the fingertips. Put crossbars or other visual identifications on all large glass doors, indicating their existence. Light will reveal the identification and prevent many bumped noses. Do not use open ceiling-mounted downlights, especially over hard or glossy surfaces—vinyl floors, furniture of polished wood, or glasstopped tables. Use fluorescent light sources in ceiling lights; incandescent bulbs need replacing too often.

how can light be supplied?

- Pendants with pulleys, which permit easy replacement of light bulbs.
- Shaded chandeliers with downlights.
- Inexpensive fluorescent strips mounted as their own shield under kitchen upper cabinets.
- Shielded wall-mounted fixtures for general room illumination.
- An oversized but underwatted ceiling fluorescent fixture.
- A whole wall washed with light in the main living area.
- As many lamps around the room as possible controlled by one switch.
- A wall bracket at either side of the bathroom mirror.
- Fluorescent chain-hung pendants.
- A fluorescent light over or an incandescent light near the tub or shower.
- An incandescent ceiling light, if it is the only choice, dimmed continually to lengthen bulb life and reduce replacements.
- For general illumination, not reading, lamps equipped with fluorescent circular tubes of 44 watts that give more light than a 75-watt incandescent or with a fluorescent screw-based bulb that give more than a 60-watt incandescent.
- A special clamp-on magnifying lamp, particularly good for small details.
- The bottom of every lampshade at eye level.
- Three-way incandescent bulbs of 50, 200, and 250 watts, permitting different quantities at different times, in lamps with a diffuser under the shade.

how much light must be supplied?

Because spaces differ in size, amount of daylight available, and interior finishes, and because fixtures differ, it is impossible to give advice regarding particular wattages. (To measure footcandles, see Chapter 2.) Generally, the Illuminating Engineering Society advises that the amount of light required to see detail and color over age 55 is at least one and a half times the amount needed for people up to 40 years of age.

- Eating dinner requires not 10 but 15 footcandles.
- Reading, ironing, desk work, laundry, cooking at the stove, and looking in a mirror requires not 20 but 30 footcandles.
- Working at the kitchen counter and painting at an easel require 75 footcandles.
- Playing the piano requires no less than 30 and possibly as high as 200 footcandles, depending upon the size and contrast of the music score.
- Hand or machine sewing and workbench tasks require up to 200 footcandles.

how much does it cost?

These lighting objectives can be accomplished with reasonably priced fixtures and lamps because the degree of improvement in lighting does not have to be related to the amount of money spent on it. Therefore, look for suitable fixtures during lighting showroom sales. Or check the prices at large department stores, especially those with catalog sales. On the other hand, most any time a throwaway fluorescent strip can be purchased for less than $10.00.

If there is any time in one's life to pour on the light, it is when one is older. The investment in lighting overall is very cheap; it pays back in living quality for the present occupant and sells well to the next. Ask any real estate agent what is the first thing he or she does when showing a residence to a prospective buyer. They turn on the lights. People are sold on well-lighted, cheerful environments. Why not make the environment cheerful for yourself or someone you know and love?

outdoor and indoor light for reassurance

15

Light outside and inside can be reassuring; it does not have to be bright, just well distributed. Outside, the larger the lighted area, the greater the surveillance potential. Therefore, illuminate large areas outside windows. In addition, illuminate all pathways—sidewalks, steps, doorways, and driveways—by pointing the light at the pathway, not at the person who might be there. Light in people's eyes blinds and hinders rather than helps them. Well-positioned outdoor light can create a pleasing, comfortable, and helpful environment.

Inside light keeps away whatever comes out in the dark and allows people to move around safely. It acts as a greeting in empty rooms and reassures with a warm glow, whether as a single nightlight in a wall receptacle or as another low-energy light source.

reassurances outside

From the inside at night, uncovered windows of any size become mirrors reflecting images and revealing only blackness beyond. In low-rise structures, the blackness can become a problem. Owners can permanently install outside floodlighting. Renter, on the other hand, must rely on temporary floodlighting. In high-rise structures, the blackness beyond the windows on high floors is the sky, not potential prowlers.

Floodlights should be shielded or protected from the weather by roof overhangs or by the fixture itself. A shielded fixture, sometimes called a bullet, needs to be 9 in. (23 cm) deep. Mount it on the structure, in the trees, or on 10- to 20-ft (3- to 6-m) high poles.

Shielded flood bulb.

Conceal floodlights in the trees.

Illuminate pleasing objects.

Illuminate pleasing objects.

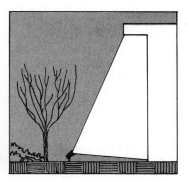

Wash the building.

Rule of Thumb for Floodlights

Floodlights will illuminate an area as large as one to two times the distance of the fixture from the ground.

Consequently, if the structure is two stories high and the floodlight is at the roofline, the light is comfortable and glare-free, projecting 17 to 34 ft (5.2 to 10.4 m) out. If the structure is one story, the light is not as glare-free and projects only 8 to 16 ft (2.4 to 4.8 m). It can be in the line of sight. Never point floodlights at driveways, walks, or entrances, as they would obscure people's vision. Likewise, do not floodlight objects that should be ignored. Illuminate objects that are pleasant to look at from the inside—trees, shrubs, walls, and fences. Lighting the outdoors beyond the windows visually expands the indoor space and makes it appear larger.

Choices for outdoor floodlights are efficient, long-life, enclosed R compact fluorescent in 15 or 20 watts (instead of 50 or 75 incandescent) or long-neck, halogen PAR in 50 or 75 watts.

On commercial structures, permanent outdoor floodlighting should wash the building's facade. It discourages vandalism and break-ins but also assists surveillance at night from law enforcement and security personnel. Further, light on the exterior of any building, whether commercial or residential, can create a favorable impression on passers-by.

Wash the facade of a commercial building with floodlights with one of these three ways:

- Floodlights on the building roof lighting down and close into the base of the structure and not aimed out more than a distance of one to two times the height of the building.
- Floodlights on the ground located a good distance away and hidden.
- Floodlights on poles some distance away and washing the facade.

In all cases, use a wide distribution of light, either by wide-beam sources or wide reflectors. The source choices are long compact fluorescent, tubular halogen, high-pressure sodium, and metal-halide. Incandescent is cheap to purchase and costly to operate. Long compacts might not operate well at low outside temperatures. Sodium puts a yellow color of light on the facade. Metal-halide operates at extreme temperatures, renders colors well, and is available in color-corrected 30 or 42K. Metal-halide is more expensive to purchase, but cheaper to operate.

Wattages from 50 to 1,500 are available to

All drawings on this page are reprinted by permission of Kim Lighting.

illuminate in any surrounding, whether dark or bright.

All commercial outdoor lighting fixtures should be indestructable, because the usual break-in method is to eliminate the light source and to enter the building in darkness. Likewise, light fixtures are sometimes beacons for pranksters to break, just for kicks.

Lighting for Entrances and Exits

Light all doorways for safe passage. Several kinds of outdoor equipment can be used:

- Wall fixtures mounted at eye level, about 5½ ft (1.7 m) high, in pairs (on either side) for major entrances or single (on the same side of the door as the lock) for other entrances.

- Wall lanterns with side brackets (overportal lanterns) above the main entrance.

- Ceiling fixtures (surface-mounted or suspended) on the underside of a flat roof overhang on one-story structures. Make sure that there is at least 7½ ft (2.3 m) clearance.

- Downlights recessed above the entrance, if the roof overhang is flat and enough plenum depth is available. Equip round fixtures with enclosed R compact fluorescent or halogen A sources; equip square fixtures with long compact fluorescent sources. If a recessed fixture is chosen for the entrance and there are wide expanses of brick or stone walls alongside, illuminate them with recessed fixtures also. Space fixtures about 10 ft (3 m) apart. Do not use them over windows. The light shines inside and might not be welcome.

People walking toward an entrance focus on it. Consequently, any glare is obnoxious. Glare is essentially uncomfortably bright light. It is created by bright direct light from the fixture and dim surrounding light. Some fixtures prevent glare by shielding and redirecting the light up and down. Some prevent glare by concealing the light with a diffuser. Some fixtures are transparent and allow the bulb to be seen. They are beautiful. Do not spoil their beauty by mak-

ing them glaring. Use only clear bulbs and make the surroundings well illuminated. Under such conditions, a clear bulb, in 25 watts or less, should not offend. Frosted or coated bulbs are not aesthetically pleasing in transparent fixtures and do not reduce the glare.

Some transparent fixtures contain downlights, adding a brighter light below. However, downlights do not give the authority to use a higher wattage above. One of the most successful uses of a transparent fixture with a downlight was in an installation I did for a client who lived in a rural area on a lake. He traveled around the country and was home only on guest-filled weekends. Part of the weekend's entertainment included a hot tub built into the deck next to the lake. His electrician had installed several security floodlights, which automatically turned on each night illuminating the deck, whether it was the weekend or not. Whenever he was at home, his neighbors would paddle by, craning their necks to see who might be in the tub. The solution was an override switch for the security lights and two wall fixtures with downlights. The top fixtures were altered to accept candles instead of bulbs, giving very soft light for hot-tub time. The bottom fixtures were left as downlights, giving bright directional light below for other times. Now the neighbors cannot see unless they paddle right up. None have done so.

Preventing Light from Spilling Into the Neighbor's Yard

Make sure that outdoor light does not infringe upon neighbors by spilling into their yard or windows. In some areas, municipal codes prohibit such infringements. In all areas, positioning outdoor lighting to keep the illumination where it belongs. Light can be confined by burying floodlights in the ground, hiding them well in trees, or mounting floodlights at least 16 ft (4.9 m) above the ground and aiming almost straight down.

Reprinted by permission of Tribune Media Services.

Bury floodlight in the ground.

Steps need light.

Walkway lighting.

Walkway lighting.

Pendants.

Post light.

Reassurance for Moving Around

Light can help people to change levels and direction safely. Light the areas for change—steps, walk and driveway junctions, and ramps—not the person. Client after client has proudly shown me his or her do-it-yourself outdoor lights aimed directly at the incoming guest. This lighting practice is an error. Navigation is hindered.

At steps, the lighting could be:

- Recessed steplight, with a compact fluorescent or end-lit fiber-optic system. Masonry steps require planning ahead; wood steps could be retrofitted.

- Recessed or surface-mounted lights on a wall adjacent to the steps.

- Overhead floodlights 10 to 20 ft (3 to 6.1 m) above with enclosed compact fluorescent or halogen PAR.

- Post light adjacent to the steps using a capsule-shaped, self-ballasted, compact fluorescent at 16 to 20 watts, equalling 60 to 75 incandescent watts. It fits into a standard socket. For emergencies, use a 60-watt incandescent signal-flasher source.

At walkways, the lighting could be:

- Compact fluorescent or incandescent fixture on a

post or stake, about 2 ft (0.6 m) above the ground. The posts should be positioned so that the size of the dark areas and the lighted areas are equal. However, if continuous light is wanted, the spaces between the posts should be around twice the height of the posts, unless the manufacturer indicates otherwise.

- Overhead floodlighting.

- Chain-hung pendants, either in regular or low-voltage, hung from a tree or an eave, lighting about a 2-ft (0.6-m) area for each foot hung above the ground.

- Wall-mounted fixtures alongside walks or steps. The fixtures should be mounted at eye level (about 5 ft 6 in., or 1.7 m) unless they interfere with moving about.

- Ceiling fixtures in roof overhangs beside walks. The minimal area lighted by this type of fixture is usually about 4 ft 6 in. (1.4 m).

At junctions, such as walks and driveways, walks and steps, streets and driveways, the lighting could be:

- Post lights mounted at eye level with well-shielded bulbs or low-wattage bulbs. Such posts can illuminate about 25 ft (7.6 m). The light from transparent fixtures should be equal

All drawings on this page are reprinted by permission of Kim Lighting.

to candlelight. Some post lights use candles for light sources; they are very distinctive for short-term illumination.

- Pole lights, such as small streetlights found in urban historic districts, lighting a greater distance. Very tall poles with security lights light even more. They can be obtained from most utility companies. They use either mercury or sodium bulbs. Think twice before using them. The color of the light is probably different from the other outdoor light. Choose them only to illuminate a large area inexpensively. They are bright. Do not install them on the bedroom side of a residence; the light is difficult to keep out. Consider the lowest wattage available; higher wattages are not necessary.

Reassurance in Parking Lots

For parking lots, the greatest amount of light over the largest area is produced by sodium lights on poles. The other choices, in order of efficiency, are metal-halide, mercury, and incandescent. Ordinarily, a parking lot that requires 400-watt mercury bulbs can use 250-watt sodium bulbs. They consume 37 percent less electricity and yield 40 percent more light than do mercury bulbs. Overall, sodium costs about half as much as mercury to operate, and mercury street lights are being replaced with these more efficient sodium bulbs. Sodium gives a golden glow similar to our image of sunshine.

Manufacturers supply technical information about the spacing of their poles, which depends upon the height of the poles and the characteristics of light distribution of the fixtures. For example, the distance from the outside edge of the parking lot to the first row of poles with one type of sodium fixture can be twice the fixture's height. The distance from an inside row of poles to another inside row should not be larger than four times the pole height. Therefore, if the poles with sodium fixtures were 20 ft (6 m) tall, the first row of poles could be up to 40 ft (12 m) from the outside edge of the parking lot. The distance from the first row to the next row of poles could be up to 80 ft (24 m).

Maintenance of parking lot fixtures is as important for long-term good lighting as is proper positioning. Clean parking lot fixtures every one to two years, and replace all bulbs every 4 years.

Automatic Reassurance

Outdoor lighting can turn itself on and off by a photosensor or a time clock. A photosensor turns the lights on automatically when the daylight gets dim—sometimes on dark rainy days. It turns them off again when the daylight gets bright—sometimes never during the winter, because the sky never gets bright enough. Unfortunately, the sensor cannot distinguish between cloudiness and sunset. Therefore, install a sensor where it will have the greatest exposure to the sky. Some posts or fixtures come with sensors; some do not. Sensors can be installed anywhere an electric wire can go.

I have done most of my outdoor lighting designs in the southeastern United States, which is heavily forested. Most of the lightposts have been overhung by trees and the sensors have been installed at the roof to gain enough exposure to the sky. Otherwise, the lights would turn on well before sunset and turn off long after sunrise.

For greater control, install a wall switch inside the building to override the signal from the sensor. The switch gives additional options. The sensor can turn on the lights after dark before anyone arrives home, and the lights can be turned off with the switch at bedtime, even though the sensor says it is still dark.

A time clock also activates and deactivates outdoor lights at predetermined times and days. However, compensations must be made for daylight savings changes by either resetting the clock or purchasing an automatically compensating clock. A time clock tends to be more wasteful of energy than a well-positioned photosensor with an override switch.

Parking lot lighting.

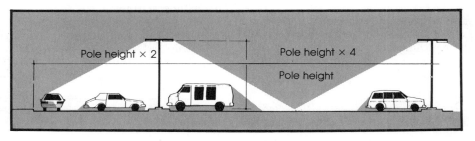

To direct emergency vehicles to a particular address, use a 60-watt incandescent source with a signal-flasher device. The device can be within the source or inserted into the socket. It is activated by switching on and off twice. It is controlled by a computer chip and sequenced by electricity. Thus, a motion-sensor switch would affect the sequence. Use without a motion sensor. The light source comes in a flame or a capsule shape.

Motion sensors can automatically reassure by lighting when motion is detected. Unfortunately, the motion of animals and temperature inversions are also detected. Overall, this is not a serious disadvantage.

Enclosed, compact-fluorescent sources do not perform well in motion-sensor fixtures. The on-off cycling shortens their 4,000-hour life.

Some light sources are protected by a Teflon coating, reducing the chance of breakage due to weather. Hot glass shatters when hit by cool water. Protect your clients!

Decisions about Outdoor Lights

Decisions about placement of outdoor lights are most important. Almost any amount of light located in the right place is sufficient, since the surroundings are typically dim or dark. Research has indicated that people feel assured if a lighted area is about the same size as the dark area ahead of them. Thus, a pathway of light, dark, light, etc., equalized would be comfortable.

The size and location of the area to be lighted govern the choice of equipment and placement. Steps away from a building require free-standing or step-integrated fixtures. Steps adjacent to a building can have wall, overhang, or step-integrated fixtures.

Light the largest area possible. It gives the best perception of brightness. Use floodlight sources reflecting light from large surfaces—walls, fences, and pavement. The amount need not be bright, unless the surface is dark in color. It must be well spread.

reassurances inside

Light inside provides reassurances. It brightens up the night. Both young and old feel more assured with a light on at night. Also, evening or overnight guests can move around an unfamiliar space undirected with the aid of low-wattage light. Such light does not have to come from a wall-plug 7-watt nightlight. Other sources can act as nightlights and provide additional functions at other times. Therefore, consider:

- A 75-watt incandescent source that slowly (20 minutes) dims down to 7 watts when turned off and on again—especially good for children's rooms.

- A 13-watt compact recessed fluorescent fixture in a bay, bow, or boxed window, over a buffet, over a hall table, or other places.

Sample Electric Cost
Used for 4 hours per night at 10¢ per kilowatt-hour, a 13-watt downlight costs 16¢ per month.

- A 15-watt decorative source in a bathroom ceiling or wall fixture. Such a fixture can identify the space for guests or customers, while enhancing the interior style.

- A 60-watt incandescent source that has an automatic turn-off device (internal or inserted into the socket) after either 10 minutes or 30 minutes on. The source blinks to warn that it is about to turn off. On/off switching deactivates the timer, leaving the light on.

- Halogen, infrared-technology PAR sources (60 or 100 watts) in down- or uplights illuminating the space between a window and sheer drapes to prevent someone outside from seeing inside. The light must be brighter than the light inside the room. This application is especially good for late-night office workers and those home alone. Owners can install permanent fixtures; renters can use tracks with cord and plug. Position fixtures close enough together to spread brightness evenly.

Sample Electric Cost
Three 60-watt halogen PAR's used in an office window 2 nights a week for 3 hours each night cost 11¢ per week at 10¢ per kilowatt-hour.

Light helps provide reassurance both inside and out. Outside, almost any amount located at the right place is sufficient. Inside, the amount can be low and be used as an additional lighting system for other reasons at other times.

facade and other floodlighting

Often businesses (and sometimes residences) want to be visible at night and have the public remember where they are. Hence, they want facade floodlighting. Well-designed facade lighting creates a favorable impression. Recommended setback for uniform facade lighting is three-quarters the height of the structure (a three-story structure 24 ft high × ²/₃ = 18 ft or 7.3 m × ²/₃ = 5.5 m). Recommended aiming is two-thirds the height of the structure (24 ft × ²/₃ = 16 ft or 7.3 × ²/₃ = 4.9). If distance must be closer, aim higher—not lower. The amount of light desirable is dependent upon the amount of ambient light. However, a general rule of thumb is that 20 footcandles is suitable.

Ground floodlighting considerations.

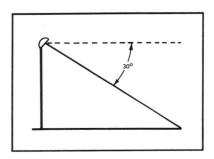

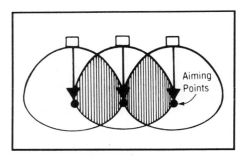

Aiming Points

Reprinted by permission of Lithonia.

Sometimes, businesses want the ground floodlighted. In general, uniform floodlighting can be from one side, from both sides or from within the lighted area. Do not space light poles farther apart in any direction than five times the mounting height. If floodlights are spaced so that the edge of their beam intersects the aiming point of the adjacent floodlight, the light will be uniform. But uniform coverage is not always necessary.

In general, ground floodlights should be aimed two-thirds out into the area to be lighted. Aiming closer to the base of the pole creates a hot spot and dimness farther out. Aiming too far out creates glare and wasted light. To avoid glare, aim floodlights 30 degrees below an imaginary horizontal line from the floodlight head. If this angle is not possible, increase the height of the pole.

Light sources for floodlighting are high-intensity discharge sources: high-pressure sodium (150 to 1000 watts), metal-halide (250 or 400 watts), and mercury (400 watts). The National Electrical Manufacturers Association (NEMA) has developed a classification for floodlights according to the beam spreads. If the source puts out asymmetrical light, two classification numbers are indicated. Thus, an asymmetrical source with horizontal spread of 109 degrees and a vertical spread of 89 degrees would be classified as a type 6 × 5 floodlight.

BEAM SPREAD CLASSIFICATION FOR FLOODLIGHTING

DEGREES	NEMA DESIGNATION
10–18°	1
18–29°	2
29–46°	3
46–70°	4
70–100°	5
100–130°	6
130°+	7

light
to create
a view
16

Structures are viewed from the outside by many people. Unless the outside light is brighter, at night the structure is primarily seen by the light coming through the windows. Thus, in residences or commercial structures where the architecture contains special windows featuring unusual shapes, the exterior impression is enhanced by the light in these windows.

bay, bow, and box windows

If windows project handsomely from a structure, bay, bow, and box windows can attract attention. The light commonly used in the room containing such a window may be sufficient to illuminate it. But if the room is not used for the major portion of the night, the special window will be unlighted. It should be lighted, nonetheless. The amount can be very small, unless, of course, the outside is lighted like the blaze of noon. The amount of illumination needed in the window to capture any attention is three times the amount outside.

Owners who are building can prewire for a recessed downlight and use a 13-watt compact fluorescent or a 27- or 30-watt reflector source. Either way, the light enhances the finish on the windowsill—stained wood or cushion for sitting. The compact has a choice of colors of light from warm to cool (30, 35, or 41K), depending on window surface colors. Choose 30K for warm color surfaces, 41K for cool, and 35K for mixed colors. At the same time, the light enhances the architecture of the window from the outside.

rose, arched, or other similar windows

Round, curved, or other specially shaped windows can be dynamic nighttime attractions that symbolize the structure. These windows can reflect light from a pale-colored surface, illuminating the window's form. The pale surface can be lined drapes, fabric stretched in a frame, or a painted plywood backing. The light source can be a fluorescent strip, cool in both color temperature (slightly bluish) and amount of heat, spreading illumination well over the reflective surface. Install two strips, one at the bottom and one at the top, or on both sides, wherever they are hidden best.

Otherwise, these windows can be lighted by reflected light from the room. If an arched window is above a door, the light from hallway

A lighted rose window symbolizes the structure.

fixtures, table lamps, or ceiling fixtures is enough. Likewise, the view of a shimmering chandelier could show through the arch and augment the beauty of both. The view of any visible fixture creates an impression of richness and high-quality ambience, desired by some and not by others. If you like to show and tell, do not miss this opportunity to discreetly flatter yourself.

clerestory windows

Clerestory windows are windows near the top of the wall, well above eye level. They can create an exterior impression and also a pattern of light on the outside. More than likely they can be illuminated by lights already in the space. However, sometimes a series of clerestory windows march across the face of a structure in various rooms. If so, make sure that all of the windows are equally and simultaneously lighted, for the best visual impact.

skylights

Skylights transmit light from the inside out as well as from the outside in. If the skylight is double-layered—a layer on the roof and a layer on the ceiling—fluorescent strips can be installed on the sides between the layers. Cover the bottom of the skylight with a translucent diffuser, letting light through but not being completely transparent. The quality of the light delivered is a moonlightlike glow.

If the skylight is a single layer only on the

A lighted clerestory window creates a pattern of light.

roof, clear incandescent bulb fixtures can make it a lighting jewel, fashioning sparkle and visual interest for a sense of vitality in the space. The light will shine above and below. Because skylights are above eye level, fixtures and clear bulbs can be used. The fixture can be an inexpensive porcelain socket or a more expensive polished brass holder. The former costs around $2; the latter costs about $16. They should hold a clear G bulb of the largest size but should have the smallest wattage possible. The fixtures should be spaced evenly along two sides of the skylight; placing them along four sides is not usually necessary.

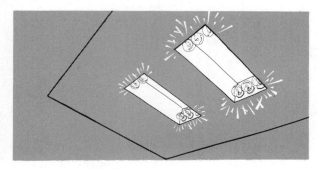

Bare bulbs create sparkle and visual interest in a skylight.

the view from inside to outside

When viewing the outside from the inside, you see either a dark void, if unlighted, or a reflection of the inside space, if outside is under lighted. Light outside must be sufficiently intense to overcome the inside brightness. To avoid a mirror reflection, inside visible sources should be hidden. All visible fixtures with bare light sources will reflect. Therefore, all sources should be indirect, recessed, or built-in architecturally. In addition, any surfaces lighted (the ceiling with indirect, the floor with recessed, and the upper-wall with a built-in cove) must not be too bright. Otherwise, the surface becomes a secondary source and reflects on the window.

Reflections obscure the outside view. The cure is to experiment with brightness levels and distribution patterns on both sides of the glass. Sometimes, more brilliance outside is needed to reduce the mirror image. Other times, a wider

distribution and dimming inside is needed. Play with it until you succeed.

One possibility for creating outside light is to install tree-mounted downlighting to filter through the leaves and create brightness on the ground in a shadowy, moon-like distribution. Such lighting is pleasant inside or out.

Another possibility for outdoor lighting is the popular low-voltage stake-light kits. They can be installed by the purchaser (owner or renter) with a screwdriver and hooked up with the convenience of a cord and plug. The package includes low-voltage waterproof cable, transformer to reduce electrical current to a shock-free 12 volts, stake-mounted fixtures, and bulbs. These systems are supplied with 18-watt bulbs. However, within the limits of the transformer, a range of other sizes—7 to 36 watts in both spots and floods—could be used for greater or lesser punches and spreads of light. The transformer's

Tree-mounted downlight
filtering through the trees
creates a moonlight-type distribution.

capacity is limited. It determines the amount of usable watts and therefore the total number of fixtures. Sometimes the fixtures are packaged with colored lenses. Resist them. Order clear ones and keep the light natural looking. (Additional possibilities for outdoor lighting are in Chapter 15.) Outdoor lighting seen from the inside has the greatest visual benefits.

lighting stained glass

Lighting stained glass is tricky. Hung in a window or as a window, it looks great from the inside in the daytime. At night, it looks great from the outside. But most people want it to look good at night from the inside, too. Frequently, outdoor floodlights are aimed at the window only to pass right through the transparent stained glass and illuminate the first opaque surface hit by the light—usually the ceiling. Alas, transparent glass does not bend light beams. Direct light passes directly through. On the other hand, reflected light reveals the stained glass. Reflected light is bent in many directions and creates a secondary source. Hence, illuminate the largest surface outside the window—a large leafy tree, an awning, a large hurricane shutter, a fence or other architectural element. When illuminated, the large surface will act like a luminous panel and let the stained glass be enjoyed at night.

When considering daylight, remember that sunlight is direct light and skylight is reflected light. Skylight is reflected from the clouds, the dust particles, and other atmospheric pollutants. The reason the stained glass is seen well from the inside during the day is that all the outdoor surfaces receiving sun- or skylight are secondary sources. They scatter the beams. The principles of light hold true for any source, man-made or natural.

Reprinted by permission of King Features Syndicate, Inc.

daylight
for a
gloomy room

17

Do you have a room that is dark and gloomy? Unfortunately, in most residences—apartments, row houses, and single-family homes—one room seems to be forgotten by the daylight. During the daytime there is not enough light to perform simple functions. Electric lights get turned on and often left on, using electricity and sending utility bills higher than needed. Electric lights were invented for use after sunset, not after sunrise. Why not get the sunlight into a gloomy room?

Daylight comes from the sun as energy in the form of light. It comes either directly from the sun or indirectly from the sky; it is free. Using it as a source of free illumination is more important today than ever. Free illumination helps to conserve our precious nonrenewable resources.

See for Yourself: Do You Have Enough Daylight?

1. Can you see the sky clearly through the window from where you are, seated in a chair while reading, eating, or working?

2. If you cannot, you do not have enough daylight to see most tasks.

3. If you cannot, identify why.

Do not consider leasing or purchasing a home or apartment without applying this test. It can save you from living in a gloomy place. Often the causes of too little daylight can be corrected. If you know the causes, you can apply the cures.

CAUSE: ARE YOUR WEATHER CONDITIONS CLOUDY OR SMOGGY?

Clouds and smog can block the sunlight. Reflected light needs to be received from the sky to get as much daylight as possible. Therefore, if you live in a cloudy climate or under smoggy conditions, use the rule of thumb for daylight.

Rule of Thumb for Enough Daylight
Have an unobstructed view of the sky from where you perform any activity that requires seeing.

Reprinted by permission of King Features Syndicate, Inc.

Clouds obscure the sun.

Can you see the sky clearly?

If not, you probably do not have enough daylight.

CURE: REARRANGE THE FURNITURE

Poorly arranged furniture can prevent you from getting sufficient daylight. Rearrangement can be the cure. Put desks, reading chairs, or any daytime work surfaces where they can take advantage of available light—but not glaring sunlight—from the window. Then daylight will be effective for illuminating whatever is needed to be done.

Desks and other tabletops can be placed parallel or perpendicular to windows.

- If you are right-handed, place the desk so the light comes from the left.
- If you are left-handed, the light should come from the right.

- If you are ambidextrous, take your choice.
- Never put your back to the window, because you cast a shadow on your work.

Place *reading chairs* so that the window light comes over the shoulder. Never face a window that could be very sunny, and never put your back to a window that is yielding very little light.

Beds can be placed at windows to gain daylight for those who like to read in bed during the day or for those who need to be in bed. For instance, does your teenager do homework on the bed with no concern for the amount of light he or she is getting? If so, put the bed at the window and have the sun supply light that nev-

When light is scarce, daytime work surfaces should be near windows.

A bed at the window permits children to read with daylight.

Incorporate electric lighting for nighttime.

er needs turning on or off. Integrate the bed to the window with a canopy headboard, enhancing the bedroom and at the same time making possible good, free light for reading. Combine draperies along the sides and a canopy board over the window as a headboard, and install a shade on the window for privacy.

Studying at night, on the other hand, requires electric light. Install an energy-efficient light with fluorescent tubes behind the canopy board for nighttime use. (Details for construction are in Chapter 5.)

In addition, you may want a pair of tall table lamps placed on either side of the bed and equipped with three-way incandescent bulbs (50, 150, 200 watts) or with fluorescent screw-based bulbs using less watts. Choose lamps with translucent shades to filter, not block, the light, thereby broadcasting it well. Or for children's rooms, double the amount of canopy lighting and omit the lamps. These arrangements are both energy-wise and foolproof.

CAUSE: ARE YOUR WINDOWS TOO SMALL OR NONEXISTENT?

For minimal daylight in any room, the window area should be equal to 10 percent of the floor area. Therefore, a 10- by 10-ft (3- by 3-m) room (100 sq ft, or 9 sq m) needs 10 sq ft (.9 sq m) of window. More area will give more light, and under cloudy or smoggy conditions, the sun's heat will probably not be too much.

CURE: ADD MORE WINDOWS

For the best daylight, consider increasing the number of windows, rather than just increasing the size of one window. Position windows on different walls. The more walls with windows, the more constant the daylight will be throughout the day. Likewise, glass in or adjacent to a door is considered a window and should be utilized whenever possible.

CURE: USE SKYLIGHTS (FOR OWNERS)

Owners can use a skylight if they have access to

Use glass in and around doors.

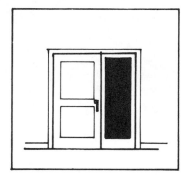

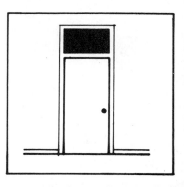

the roof. Skylights are more than just holes in the ceiling. They are windows to the changing panorama of the sky—fast-moving clouds, tree-tops, changing sunlight and, at night, a handful of stars and the moon. Skylights gather all the light possible. They are made of clear or non-clear glass or plastic. Some are equipped with interior screens, exterior awnings, built-in roll-er shades, or ventilation flaps. Some are domed; some are flat. Some are a single layer; some are double and sealed.

The general tips for skylights are:

• On sloped roofs, position skylights preferably on the north slope of the roof. Since skylights gather so much light, north light is softer—and better. Artists have known this fact for many years. If they are not possible on the north slope, then, in order of preference, east, south, and west are suitable. Skylights sloped to the south act as solar heaters, warming the air.

• On flat roofs with clear skylights, install a shade or other device to shut off the sun when it becomes too hot or too glaring.

• Choose a nonclear skylight for rooms that contain carpeting, wallpaper, and fabrics that might fade, particularly in blues and purples.

• Choose a clear skylight for rooms that do not have delicately colored finish materials or art-work subject to fading, such as kitchens, dining areas, atriums, hallways, bathrooms, and en-closed porches (Florida rooms), and any living area with mostly hard surfaces.

• Use nonclear skylights when the view should be obscured; for instance, on flat roofs where leaves will pile up.

• Use nonclear skylights to soften too much direct sunlight.

• Choose a nonclear skylight where the bright-ness impact on the ceiling needs to be minimized.

• A clear skylight can be finished by boxing in the space between the ceiling and the roof. The box-ed-in part will be as deep as the distance be-tween the finished ceiling and the roof.

• Those people that do not like a deep boxed-in effect should finish off the ceiling with a flat nonclear plastic or glass diffuser, not interrupt-ing the surface of the ceiling.

Specific tips for clear skylights are:

• Even in cloudy or smoggy areas, some days are sunny. On these days, sunlight through clear skylights can cause blinding brightness and in-

A clear skylight is usually recessed in the ceiling.

A diffuser on the ceiling eliminates the boxed-in look.

terfere with activities. Position the skylight so that the direct sunlight will be where you want it and when you are most likely to be in the room.

- In a kitchen, a skylight toward the west will admit glaring afternoon sun, just at the time the evening meal is being cooked. A skylight facing east or north would receive only reflected light in the afternoon. If a western orientation is the only choice, use nonclear glass or plastic to soften the light.
- In bathrooms, a skylight toward the east might thrust glare on the mirror in the morning. Make sure the sun's rays cannot reach the mirror. Strong, but not glaring light is usually welcomed in the bath.
- In bedrooms, a skylight toward the north is best. Toward the east will bring in early morning light that might disturb your sleep. South or west exposures can bring in strong sunlight that can easily fade your carpet or other materials and heat up the space too much in the summer.

In spite of the cautions, clear skylights add a visual dimension to a room not possible with a window. The passing clouds and the moon can be seen through them. On cloudless nights, moonlight is bright enough to allow you to move

around without turning on any electric lights. In the winter, moonlit nights are brighter than early mornings.

CURE: INSTALL GLASS DOORS

Usually owners do not have access through to the roof for a skylight. However, if owners have a solid door in the room that faces a well-lighted space (interior or exterior), they can replace the solid door with a glass door. The glass door acts like a window, provided it is opposite brightly lighted windows in another room (windows facing south, preferably); it is in a bright, sky-lighted hallway; or it is an outside door.

There are more varieties of glass doors, including folding glass doors, than there are varieties of wood doors. Most glass doors have at least a wood or a metal frame. Do not forget the all-glass door, usually used in commercial buildings, but nonetheless suitable for residences. It is sophisticated-looking. An all-glass door becomes a full-length window. Privacy, if needed, can be maintained by choosing patterned rather than clear glass. It is a successful solution for rooms that do not have any windows.

Renters are usually unable to change the structure they lease, by virtue or vice of their property rights—the right to live in but not to alter without permission. Renters must rely for the most part on internal changes to augment daylight. (Review the sections on furniture arrangements and interior surfaces in this chapter.) However, a wooden door could, for the term of the lease, be replaced with the renter's wood and glass door, if it is hinged in the same way. Then, the renter could enjoy the daylight and reinstall the solid door before moving out.

CAUSE: IS THE SUNSHINE ON THE OTHER SIDE OF THE HOUSE?

CURE: CREATE EXTERIOR ARCHITECTURAL SURFACES

If you are the owner, create a large exterior surface such as a fence, wall, or a patio floor to redirect the sunlight into your room. Exterior surfaces, such as patios, pavement, the ground, fences, and walls, can reflect almost half the light they receive through the windows into the adjacent room. On sunny days the amount of direct sunlight striking these surfaces is more than that needed for interior use. On cloudy days the light reflected from the ground can be brighter than the sky itself.

Replace a solid door with a glass door to borrow light from another space.

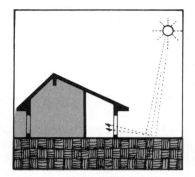

Light can be reflected.

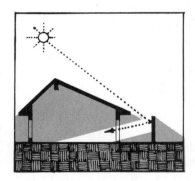

A fence can redirect the light.

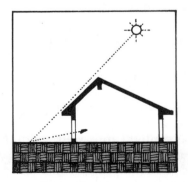

Create a surface to reflect light.

Is the sun on the other side of the house?

A concrete patio surface reflects light through the windows.

Rules of Thumb for Exterior Reflecting Surfaces

The closer to the interior space, the brighter the light; the farther away, the softer.

The more white in the color of the surface, the brighter the light; the more black, the gloomier the light.

Be aware that pure or off-white (highly reflective colors) on these surfaces might produce glare. Pale to medium color values are reflective enough. Select the color among these values. Light will be redirected and usable. For example, a fence stained a rosy tan would reflect 30 percent of the light. A beige-colored brick patio would reflect at least 48 percent of the light back into the adjacent room. Both have pale to medium values.

GROUND REFLECTING MATERIALS

TYPE	AVERAGE REFLECTANCE
Natural Minerals	
pebbles	13
sand	30
water	95
bare ground	7
slate	8
Natural Vegetation	
dark green grass	6
other ground covers, ivy, etc.	25
Man-made Materials	
concrete	40
asphalt	7
brick—pale tones	48
brick—dark tones	30
wood deck	40
plastic grass	45
white terrazo	45

CURE: CREATE INTERIOR SURFACES

Colors on the interior surfaces can cause a room to be gloomy, and also they can cure it. Renters and owners can enhance the daylight by creating interior surfaces that redirect the light. Walls, floors, and ceilings become the most important control available. They need to be finished in colors that reflect as much light as possible. Pale colors with large amounts of white reflect as much as 85 percent of the light within the room. Dark colors, on the other hand, absorb so much that sometimes less than 10 percent is reflected.

For a gloomy room, the back wall (the one opposite and usually the farthest from the window) is particularly important. It adds or detracts from the total room illumination. A back

On gloomy days and at night, use an energy-efficient cornice light on the back wall.

wall receives little direct light, even on sunny days. The light it receives is already reflected from somewhere else. If the color on the back wall is highly reflective, the whole room can feel bright and airy. To increase the apparent light, use pale colors, mirrors, metallic wallpaper, or polished tile on the back wall. Create reflected light while making an interior design statement.

In addition, for very gloomy days or nighttime, equip the back wall from end to end with a fluorescent cornice light. The cornice will create a cheerful atmosphere by spreading the energy-efficient light where needed. It will heighten the colors and texture of the wall covering and enhance nearby furniture. (Review cornice light in Chapter 5.)

Never paint a ceiling in a gloomy room a dark color; do it only in a sunny room. Ceilings send back the light that is reflected from the ground, and gloomy rooms need all the light they can get.

The floor receives light from the sky, so keep the floor covering as pale in color as possible, reflecting back as much light as is available.

Other walls should conform to the average recommended for interior surfaces.

Rule of Thumb for Average Reflectances of Interior Surfaces

The average reflectance of interior surfaces should be between 45 and 60 percent.

A wide choice of colors is contained within these minimums and maximums. This range does not limit the color treatment of surfaces, particularly walls, to white alone. Pale values of red and green, for instance, are comparable to the values of off-white in light-reflecting ability. Do not fixate on only white.

How do you know what the light reflecting value of a color is? Technical paint samples list the values as light reflectance percentage (LR%). The percentage is light thrown back; the percentage missing is light absorbed. Such color samples can be compared to the color of wallpaper, fabric, furniture, and other interior surfaces, indicating approximate light reflectance.

However, without the samples, color reflectance values of common items may give you some measure for judging colors. The color of a ripe tomato has a light reflectance value of 25 percent; the color of pine needles, 20 percent; the color of French vanilla ice cream, 80 percent; the color of butterscotch sauce, 60 percent; and the color of leaves on the trees in spring, 40 percent. If the color you are considering is about the same as your morning coffee grounds—13 percent—it is too dark.

MINIMUM AND MAXIMUM REFLECTANCE VALUES
FOR INTERIOR SURFACES
FOR GAINING DAYLIGHT IN INTERIORS

	MINIMUM	MAXIMUM
External		
ground	20	70
vertical surfaces	25	40
Internal		
ceilings	80	95
other walls	40	90
floor	20	70

CAUSE: ARE EXTERNAL ARCHITECTURAL OBSTRUCTIONS BLOCKING THE DAYLIGHT?

Architectural obstructions that block the daylight are usually adjacent buildings, fences, walls, and large roof overhangs.

CURE: ALTER THEM IN SOME WAY

Owners who have external obstructions can alter them if they own them. If the obstruction is

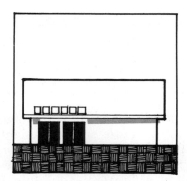

Cut out a roof overhang.

removable—a roof overhang—cut it back or open it up over the window, admitting light. If the obstruction is not removable—a building, a fence, or a wall—paint it a value that reflects light—a color with a lot of white in it.

Renters who have external architectural obstructions cannot alter them structurally but might be able to repaint. Also, renters can rearrange their furniture and cover their interior surfaces with highly reflected colors to gain all the light possible.

CAUSE: ARE EXTERNAL LANDSCAPE MATERIALS BLOCKING THE DAYLIGHT?

Landscape materials that obstruct are large trees, overgrown shrubs, and vines climbing on the structure.

Pruning a dense tree . . .

. . . permits light to enter.

CURE: ALTER, REMOVE, OR REPLACE THEM

Alter landscape materials judiciously. Reshape but do not disfigure them by pruning. Remove or replace those that cannot be pruned satisfactorily. Replace evergreen materials with deciduous materials that drop their leaves in the winter and grow leaves in the summer. Replace broad, dense trees and shrubs with open, smaller-leafed ones. They give airy shade and some light. A Japanese maple, for instance, gives soft shade, drops its leaves in the winter, and is an attractive shade tree, not too dense, as compared to a sugar maple, which is broad-headed, large-leafed, and very dense.

Owners who cannot alter landscape materials need to consider using skylights or glass doors and altering exterior architectural surfaces as suggested in this chapter to enhance the daylight. Renters who cannot alter the landscape materials need to use furniture arrangements and interior surface changes included in this chapter. However, if the landlord pays the electric bills, it might be possible to motivate him or her to assist in getting good daylight with other cures, so that the cost of utilities could be reduced.

keep the
hot sun out
but let
the light in
18

Even though sunshine makes us feel good, too much sunlight in our interior spaces can have adverse effects. It can bring in too much heat and be too glaring when it is direct. Sometimes it restricts the use of furniture and activities, such as preventing people from sitting in a particular chair or on one side of a table. Likewise, sunlight can fade upholstery fabric, blister furniture, and bleach carpets.

Can the light be kept and the heat and harshness be shut out? Yes, the sun's light can be obscured or eluded. Obscure it by blocking or scattering the light by means of window treatments, architectural devices, or landscape materials. Obscuring methods can be applied both indoors and out. Elude it by getting away from the direct light by means of furniture arrangements or dark interior finishes. Eluding methods must be applied indoors.

choosing a sun-control method

What is your status—owner or renter? If the space or the structure is not owned, it cannot be altered, at least not without permission.

What is the height of the structure? Outdoor obscuring devices—trees, fences, and others—can be used with a one- to three-floor structure (low-rise). They cannot be used with a four or more floor (high-rise) structure, unless the device originates from a balcony (potted tree).

What is the location of the sun? If you know where the sun is at various times of the day, you will know where to expect direct sunlight. In the early morning the sun is low in the sky, rising to the highest elevation at noon standard time, descending again and setting at the end of the day.

All the time the sun is rising and descending, it moves from an easterly direction in the morning to the south at noon, to a westerly direction late in the day. The arc of that movement is larger in the summer than in the winter. In the summer, the sun actually rises north-northeast and sets north-northwest; in the winter it rises east-southeast and sets west-southwest. Consider this information in relationship to the windows in the room.

What is the angle of the sun? Knowing the angle above the horizon at various times of the day permits you to determine the position for a sun-control device. The sun is at low to medium angles in the east and west. The sun is the high-

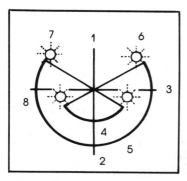
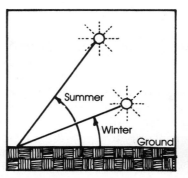
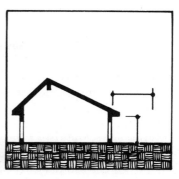

Sun's arc (1) N, (2) S, (3) E,
(4) winter arc (5) summer arc,
(6) rise, (7) set, (8) W

Angle of the sun.

How high and how wide should
a sun control be?

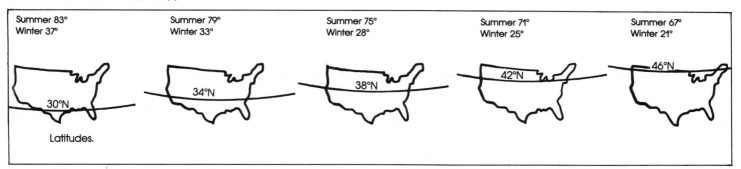

| Summer 83° | Summer 79° | Summer 75° | Summer 71° | Summer 67° |
| Winter 37° | Winter 33° | Winter 28° | Winter 25° | Winter 21° |

30°N

34°N

38°N

42°N

46°N

Latitudes.

Sun's altitudes at noon

est in the south at noon each day, but the angle is different for each latitude. To determine the angle of the sun, know your latitude. In lower latitudes the sun is higher in the sky; in higher latitudes the sun is lower.

At the same time, the sun is the highest in the sky in the summer, whatever the latitude, and lowest in the winter. All in all, it is easy to remember that the sun is highest at noon in the summer at south windows, and at the east and west windows it is essentially low in all seasons.

Do not try to build, buy, or rent a sun-control device before you know three pieces of information: how high a sun-control device should be, how far it needs to stick out, and what the most effective position is. Determine this information by pretesting it with a pencil and paper.

Pretest Sun-control Devices Yourself

1. Determine the nearest latitude. If you are halfway between two latitudes, divide the difference in half, adding the half to the lower number. Thus, if you live in Philadelphia which is halfway between latitudes 38° and 42°, the summer angle would be 67½° and the winter angle would be 24°.

2. Make a small-scale drawing of the room and the surrounding walls, like a drawing of a doll's house cut in half, and a drawing of the device being tested—a fence, a tree, an overhang.

3. Using a protractor, put the center point at the edge of the sun-control device—the edge of the roof overhang, the edge of an awning, or the lower and upper edge of a tree.

4. Project the angle of the sun at your latitude for both summer and winter, directly into the room. Everything on the far side of the line from the sun would be in shade; everything on the near side of the line toward the sun would be in the sunlight.

Use a protractor . . .

. . . to test a sun-control device.

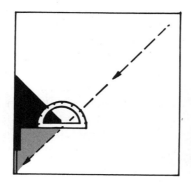

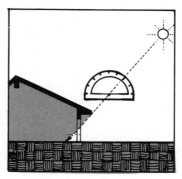

This pretest will show the summer's sun penetration (the least) and the winter's sun penetration (the most). The effectiveness of a control device can be determined before investing in it. Is a trellis over a window going to give enough additional shade? Is a medium-sized tree tall enough? Who wants to build a deck roof only to find that it is too short to give shade? Likewise, this kind of information permits pretesting eluding methods. It is truly a handy tool to ensure that money is spent wisely and effort is spent efficiently.

What is the type of activity in the room? Sedentary activities are more likely to be disturbed by harsh sunlight than nonsedentary ones. For instance, sitting at a desk and reading social security forms in harsh, hot sunlight would be more uncomfortable than taking inventory of the supply cabinet. Likewise, many short-term activities, such as brushing teeth, are rarely affected by too much sun. Therefore, sun control is not as necessary. On the other hand, many long-term activities are affected by too much sun and do require sun control.

window locations

Windows are the only way that sunlight can enter a structure. Their location is important, and the direction they face determines when they receive sunlight.

Windows facing the east receive direct light in the morning at a very low angle. Devices are necessary for bedrooms for those whose sleep would be disturbed by light or for offices where people get to work early.

Likewise, west windows receive the sun directly in the afternoon in a decreasing angle till sunset. At that time of day it seems harsher and brighter. Therefore, sun-control devices are necessary for dining rooms so that dinner and glaring light do not come together, and for offices where late afternoon work might be impeded.

South windows receive sunlight between 10:00 A.M. and 2:00 P.M. In the summer, sun control is obtainable by outside devices that block most, but not all, the sun and reflect some light. In the winter, sun control is sometimes needed, but since the sun warms it is often welcome. Nonetheless, it is lower in the sky and penetrates deeper into the space. When heat is not wanted, daytime activities can be lighted by the sun, as long as sun controls are available.

North windows never receive direct sunlight. The light coming in a north window is reflected from somewhere else—the sky, the ground, or another building. North windows do not need any obscuring device, not even a roof overhang. More often than not, however, structures are designed with the same overhang all around. This practice ignores the sun's arc through the sky and indicates a need to equalize the roof—probably unnecessary.

How many windows are in a space and where they are located are important. If only one wall has a window, the control device for that window governs the sunlight for the whole space. If the device is heavy draperies and they are closed, the direct sunlight gets shut out along with all the reflected skylight. Frequently, after the drapes are closed, electric lights are turned on to allow everyone to see, even during the daytime. Therefore, the sun is controlled, but money is spent on electric light. If daylight is available, use it. In general, small rooms can manage with only one window for light, but larger rooms need additional walls with windows to get sun at different times of the day.

More control is possible if two walls have windows. When one window receives direct light, it can be blocked and the other window can provide reflected light. More often than not, people seated in such a room are likely to face one of the windows, and therefore some form of sun control is necessary. Moreover, sometime during the day, at least one of the windows will receive direct sunlight.

Similarly, rooms with windows on three or four walls require sun-control devices. At least two or three windows will receive sunlight sometime during the day. Therefore, any and all control devices are appropriate. For greater control, use several devices at the same time. (I have had only one lighting client that had a room with windows on all four walls; the space was spectacular, but the client was unhappy because the sun faded her wall-hung art. We needed to obscure the sun and not put in any electric lights that would fade the art further.)

Clerestory windows (windows on the wall above 8 ft, or 2.4 m) give good light, but the direct sunlight can penetrate and fill the room with glare. Indoor controls are difficult to reach but not impossible if well engineered to be managed from the floor. Light can also be controlled with outdoor devices and ceiling color.

Skylights gather the greatest amount of sunlight possible, spreading it throughout the room. Like clerestory windows, indoor controls are difficult to reach unless they are well-designed. Therefore, consider using a skylight with a built-in louver or purchase a motorized shade to control the sun. If neither of these controls is available, drape a cloth or use a large umbrella, some wooden baffles, or some landscape materials, inside or out, to soften the light. Never install a skylight without sun control in a room with surfaces that are prone to fading (draperies, carpet, and wallpaper). Instead, install in a room that does not require light control, such as a bathroom, hallway, kitchen, or any space that has mostly hard surfaces—tile floors, wicker furniture, and painted walls.

methods of control

Obscuring

Control the sun by obscuring it—blocking or scattering the light. Obscuring methods are window treatments, architectural devices, and landscape materials.

WINDOW TREATMENTS

A window treatment is putting something at a window to reduce the sun's glare and harshness. However, it will not reduce the heat unless it is insulated. A highly adjustable window treatment permits the light from any angle to be redirected, scattering and softening it. They are best for sun control. Each type of window treatment has different degrees of adjustability and control the sun at different angles.

Draperies are the most common window treatments. But they are poor controls if windows are only on one wall. Depending upon the thickness of the material, draperies can either block or scatter the light. Dense material blocks; thin material scatters. When they are fully closed, dense draperies block the sun low in the sky. When they are partially closed, they sometimes block the sun at oblique angles. Thin drapes cannot block the direct sun at all. If thin drapes cover the window at all times, make sure that electric light is not also being used during the day. Daylight is free, and it should be controlled in some other way.

Grilles, screens, and latticework scatter the sunlight coming from an angle, but in most cases they cannot be adjusted to control the low sun. This treatment works best when the sun is high.

Shades can block all or part of a window. When part of the window is blocked, the rest of it transmits light from the sky or ground. To accomplish control, shades can be pulled from the top or the bottom. If pulled from the top, they block the sun high in the sky; if pulled from the bottom, they block the low sun. Therefore, aesthetics or convenience alone should not determine which way to pull shades. Sun control should help.

Shutters may be louvered or solid. Louvered shutters can adjust like venetian blinds and scatter the light. Solid shutters block the light completely or partially, depending upon their position.

Venetian blinds are the most adaptable of interior window treatments. They can deflect sunlight and redirect it to illuminate the room. Light can be scattered up to the ceiling, down to the floor, or to the left and right, depending on whether the blinds are horizontal or vertical and how they are adjusted. Whatever the adjustment, blinds are the most versatile sun-control device. (Window manufacturers in Scandinavia make windows with the venetian blinds built in between the layers, which makes them dust-free.) Again, aesthetics alone should not dictate whether the blinds should be vertical or horizontal. The decision should include control of light.

The color of a window treatment affects the color of the light coming in through the window. Light picks up the tint of the object it strikes. Of course, it strikes the window treatment first.

See for Yourself: Does Light Pick Up Color?
1. Cut the bottom out of a shoe box.
2. Tape waxed paper over the bottom (not clear plas-

tic food-storage wrap but old-fashioned, semi-transparent waxed paper).

3. Shine a flashlight inside the box, aimed at one side.

4. In front of a mirror, observe the color of the light seen through the waxed paper.

5. Lay two pieces of bright-colored paper (red, orange, or deep purple construction, Christmas, or other kind of paper) inside the box along the sides.

6. Again shine the flashlight on one side and observe the color of the light. It will be tinted with the color of the paper you have put inside the box.

The best color for a window treatment is neutral or near neutral; the worst is an intense color that distracts or becomes a source of glare itself. Choose a color that gives no unwanted effect of its own to the interior space.

The construction of a window is very important because it creates sun control. Ideally, sun control should be accomplished by structural gradations from the brightness of the outside to the dimness of the inside. Windows in older homes created such a gradation. They were embedded in thick walls. They had deep sills, tapered wood between the glass panes, and movable shutters both inside and out. On the other hand, in contemporary buildings, windows are placed in thin walls, offering little to soften the light, especially in curtain-wall office buildings. Special attention must be paid to deliberately softening the light at and around windows. Softening can be achieved in contemporary construction by using window treatments and landscape devices as well as adding other architectural devices.

ARCHITECTURAL DEVICES

Architectural devices block and scatter the sunlight. They can be adjacent to, attached to, or part of the structure. They can be stationary, movable, or even removable. They give complete or partial shade, reducing the harshness of the sun for the inside spaces. In addition, heat is prevented from ever reaching the structure by devices that block the sun's rays. In hot weather, an air conditioner or fan does not need to remove the additional heat, and money is saved. External architectural devices can obscure the sun in different positions in the sky, but for the most part they are not adjustable.

Awnings can be either adjustable or nonadjustable. They can have side panels, blocking the sun from high, medium, and low oblique angles. Awning material has shorter life expectancy than the structure to which it is attached.

But they are decorative and add to the ambience, besides being functional.

Balconies block the sun for the window below, thereby becoming a roof overhang. Apartment dwellers in structures with balconies appreciate their upstairs neighbor's balcony more days of the year than the neighbor.

Fences, walls, courtyards, and other buildings block the sun, particularly the rising and setting sun. Moreover, they can provide privacy and reduce noise. The architecture of fences, walls, and courtyards can be visually, auditorially, and socially pleasing, while it provides sun control.

Sun protection and privacy from a fence.

Patterned glass and sunshielding glass or plastic cut down the sunlight in different amounts. Patterned glass diffuses the light unless the sun is aiming directly at it. Then, the light appears brighter than through clear glass, because the pattern augments and makes the light brilliant in the way that cut glass does. (Patterned glass is most often used in shower stalls.) On the other hand, sunshielding glass or plastic is either reflective or tinted. Reflective glass or plastic bounces back between 8 and 80 percent of the sun's light, depending upon its manufactured characteristics. Tinted glass or plastic bounces back 6 to 8 percent. However, direct sunlight is a problem for these sunshields, too. Question your supplier carefully to determine the exact amount and exact type of sun control you are purchasing. In high-rises, where other architectural options for suncontrol are limited, sunshielding glass or plastic is a must; in low-rises it can be very useful.

Latticework is a decorative form of crisscrossed wood or metal strips, which can be used to cover a window. Screens and grilles are usually metal. All three can diffuse the light somewhat, but none can block the direct sun, because of the openness of their construction. Scattering the light, in some cases, might be enough.

Louvers can be added to the outside of many structures to block the sun at the angle needed. They are suitable for most building styles, except traditional period styles like Neoclassical, Southern Colonial, and Cape Cod. Louvers have been used successfully for many years on residences in semitropical climates and on contemporary commercial structures in all climates.

Porch or deck roofs block the sun not only for the porch or deck but also for the adjacent interior space—two benefits from one device.

Recessed windows create a gradual gradation of light from the sky to the room.

Bay and bow windows function as recessed windows if the depth of the bay or bow is great enough to shade the interior space from direct sunlight.

Roof overhangs are permanent devices to block the sun as well as to keep the rain off the windows. On the east and west sides, overhangs cannot block the sun because it is usually too low in the sky. On the north, an overhang is not needed for sun control. (The sun never shines directly in north windows in this hemisphere.) On the south, overhangs are beneficial.

Shutters can be used on the sides of or over windows. Either way, they must be movable to be effective. Ideally, side window shutters should close completely and be controlled from the inside. Overhead window shutters can close completely but unfortunately can never be completely opened. Consequently, they always block the light from the sky but allow reflected ground light. Without a doubt, all fixed shutters are useless for sun control and are strictly decorative.

Trellises are cross-barred metal or woodwork. Trellises can be placed at right angles alongside the window or over the top. Trellises along the sides diffuse the light from an oblique angle (the rising or setting sun). Above the window, trellises diffuse the sun when it is high in the sky. A trellis can support vines or other landscape materials, adding a rich, softening effect besides giving the bonuses of color, smell, and beauty.

Which architectural devices are usable are determined by the type of structure and whether that structure is owned or rented. For example:

- Roof overhangs are not applicable for high-rises, unless you are the owner or builder and are able to specify overhangs on the building.
- Trellises, awnings, lattice, removable grilles,

and louvers are excellent devices for renters of low-rises. They can be built, installed, and removed when moving time comes around.

- On the other hand, a courtyard, fence, or other building could be built by owners in low-rise structures to obscure western setting sun.
- Movable shutters are usable for owners, especially those who plan to build.

LANDSCAPE MATERIALS

Landscape materials—trees, shrubs, vines, and hanging plants—can block or scatter direct sunlight. These materials can be movable (a small tree in a tub on wheels), or they can be permanent for the life of the material (a planted

Tall trees obscure . . .   . . . the sun at mid-day.

hedge). Some materials always produce shade (evergreen); others produce shade only in the growing season (deciduous). Determine the height of landscape materials and choose those that can obscure the sun now and later. Small trees at eye level (around 5 ft 6 in. or 1.7 m) obscure the sun very low in the sky when rising and setting. Medium-sized trees (around 20 ft or 6 m tall) obscure it before midday in the summertime and during midday in the winter when the sun is lower in the sky. Tall trees obscure the sun high in the sky, at midday in the summer.

The broader the landscape material, the wider the area it shades. The denser the growth, the more sun it blocks. For instance, a small-headed tree shades only a very small area. A broad tree shades a larger area. In addition to providing shade, landscape materials can also provide privacy, enhancing your indoor and outdoor living.

At different times of the day, different types of landscape materials are best for obscuring the sun. Early morning and late afternoon sun protection can be gained by hedges and shrubs that grow compactly from the ground and are high enough to obscure the sun. Middle of the day protection can be gained by medium to tall trees with broad, compact growth, by

Broad landscape materials . . .   . . . block the sun.

Rules of Thumb for Furniture Arrangements

Work surfaces, tabletops, and desks should not be next to windows that receive direct sunlight at the time of day when the surfaces would be used.

Sofas, chairs, and lounges should not face bright, glaring windows, particularly if the seat is bathed in the sun.

In a low-rise structure, furniture arrangements for eating should not require a seat to face an unprotected east window (at breakfast time) or an unprotected west window (at dinnertime). In a high-rise, the sun would probably be below the level of the west window at dinnertime.

vines or other plants growing on a lattice and hanging baskets. Tall trees also can protect high clerestory windows from the noontime sun.

Get a bonus by choosing landscape materials that produce flowers, particularly fragrant ones. For the fullest enjoyment, choose those landscape materials that bloom when the climate allows you to be outside or to have your windows open.

Unlike other devices, landscape material is living and has specific requirements for growth—light, soil, and moisture. Choose the materials whose requirements you can satisfy and whose susceptibility to disease is low. You will want to ensure that the device will be reliable over the years and will not need to be replaced.

Eluding

The sun can be controlled by getting away from its light. One method of eluding is to arrange the furniture away from the sunlight. The second method is to cover the major surfaces of the room with colors that either absorb some of the sunlight or blend with the bright light.

FURNITURE ARRANGEMENTS

Furniture arrangements are often the last method employed for eluding sunlight, but they should be used more often. In rooms facing south windows, use two different furniture arrangements—one for summer and one for winter. In the summer, the furniture can be closer to the windows but not close enough to be in the sun. In the winter, the furniture should be farther away, since the sun penetrates more deeply. With any arrangement or any window exposure, several rules of thumb should be followed to elude the sunlight.

INTERIOR FINISHES

All major interior surfaces—floor, walls, and ceiling—are finished in a color. Sunlight striking these surfaces will be absorbed and reflected back in the quantity allowable by the color of that surface. Dark colors (dark blue, black, deep brown) reflect only a small percentage of the light. Pale colors (beige, white, pale yellow) reflect most of the light received.

The floor receives most of the direct sunlight in any room. To reduce the amount of light, the floor can be made dark with wood stain, carpet, or other floor covering.

The wall containing the window is the most critical wall for eye comfort. If the window receives direct, unrelenting sunshine, this wall should not be dark. It should be pale so that no harsh contrasts are created.

Contrary to the usual practice, pure white walls should not be used where windows receive unrelenting sunlight. In sunny climates and in high-rises, windows are more likely to receive direct sunlight for long periods. Consequently, pure white on the wall is too bright. White creates an unpleasant brightness inside that does not subdue the outside brightness. Instead, use a healthy shade of pastel or a color that reflects only 70 percent of the light. There are many shades and colors to choose from in this reflectance range. How do you know how much a color reflects? Ask your local paint store for technical paint samples. These samples list light reflectance values, indicated by LR%. The percentage refers to the amount of light reflected; the amount remaining is the light absorbed.

Walls adjacent to the window receive sunlight. If the walls are dark, they become harsh contrasts. Finish them with a high light-reflecting color, but not necessarily white.

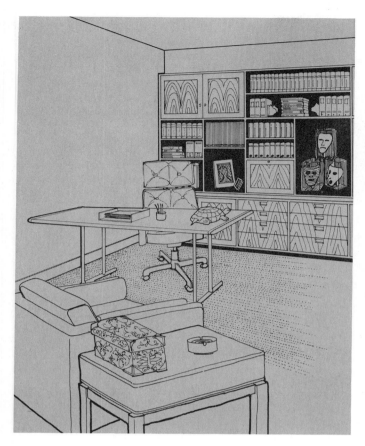

In sunny climates, put daytime work surfaces away from windows.

A dark window wall creates too much contrast in sunny climates.

The back wall of a room can be finished in a dark color. It does not contrast with the bright sunshine and helps to absorb excess light.

Ceilings also can be dark to absorb light, since direct sunlight reaches the ceiling only very early in the morning or very late in the day. Dark ceilings help control the light from clerestory windows if no other control device is possible.

FURNITURE COLORS

Any furniture, particularly a desktop or other work surface that receives direct sunlight, should be finished in a nonglossy, pale color to avoid eye-fatiguing contrast. But furniture away from the direct sunlight can be dark, thereby absorbing some light. (Mediterranean countries use dark furniture; for example, Greek cottage and country Italian furniture.)

using multiple-control devices

Sun-control devices should be used together. They can effectively moderate the light throughout the changes of the day, the months, and the year. They can work either simultaneously or sequentially.

Multiple devices work simultaneously to soften the light. For example:

Outside
- tree above the roof
- trellis adjacent to the window
- shutters and hanging plants inside the windows

Inside
- window wall finished in a pale color

Multiple devices.

Simultaneous devices sometimes block and sometimes filter. The usable light is kept. Heat and harshness get shut out.

Multiple devices work sequentially, controlling the sun in its daily arc through the sky. For example, a family room, with windows facing east, south, and west and used all day long by a member of the family, could benefit from several control devices.

Outside Devices

- fence along the northeast and southeast lot lines
- group of small dense trees near the windows at the east to southeast
- awning on the south facing sliding glass door
- group of medium-tall dense trees and dense shrubs located away from the windows to the southwest
- carport on the west side

Inside Devices

- vertical venetian blinds at all windows
- ceiling painted a dark color
- furniture arranged to receive the best reflected light

Together these sun-control devices would regulate the sun for a single-family, owned home in a very sunny climate. Other types of structures, ownership, and climates would require a different set of devices.

the
best possible
desklight
19

Office design work used to be simple. The worker's tools were a desk, a chair, a telephone, a wastebasket, and a file cabinet. The office was in a commercial building. No longer. Computers and electronic transmission have changed all that. Offices are also at home. Now, a modem, a fax, and a computer are additional tools. Files are on discs. Further, computers, televisions, and other self-illuminated screens are in the office—a video screen on the telephone, and a touch screen for automated controls. At home or at the corporation, the issue is productivity, productivity, productivity. The lighting designer must design light-ing to enhance the required productivity. The job has become more complex. Visual tasks are now paper-based and screen-based. Unfortunately, each task needs a different lighting solution. The best light for a paper-based task can inhibit a screen-based task. Since 85 percent of the infor-mation received at a desk is transmitted through the eyes, desk lighting must maximize the ability to see. Whether the desk is at home or at the cor-poration, make it the best possible desk light. The best possible is the best amount, the best color, and from the best position to enhance vision, and provide good visibility.

the best amount

The amount of light needed is influenced by the task, the viewer, and the surroundings, and could be regulated by a governmental entity. Often states or local government restrict the amount of light for commercial offices. Rarely do they re-strict the amount of light for home offices. The restrictions are aimed at reducing energy con-sumption in order to decrease greenhouse gases and acid-rain emissions. Be sure that the restric-tions are fulfilled in the jurisdiction of the office being designed. Remember that the strictest codes take precedence over less strict codes no matter what the jurisdiction.

The Task

The task is analyzed by the readability and the time available to read. Readability includes task size, color, and contrast. On the one hand, pa-per-based tasks might be poor photocopies with paper and print two shades of gray. Or, it might be small-sized, but important, numbers being put into a computer. Are glossy technical jour-nals the reading materials? Are penciled long-hand notes required reading? If these or other difficult paper-based tasks are being performed, more light from the right direction is needed than if the tasks were sharp, simple, and unimportant. And long-term tasks require the best lighting.

On the other hand, screen-based tasks are self-illuminated at various illumination levels. For example, a notebook computer with an LCD screen is dim and appears dimmer with high room luminance. The image color and background color of a self-illuminated screen are critical. White or pale-colored screens with dark images can withstand higher general room luminance than dark or deep-colored screens with white images. Thus, the amount of light in the room needs to be tailored to the color of the self-illuminated screen.

Likewise, a screen acts like a mirror reflecting any lighted surface (a shirt) and any visible lighted source (a ceiling luminaire). Therefore, what the screen "sees" affects the visibility.

If the task is a video phone and the caller's image is also being transmitted, the room lighting needs to be like a teleconference space and surrounding luminance needs to be controlled.

The Viewer

The amount of light needed is also influenced by the viewer. The viewer's visual acuity, relative contrast and luminance sensitivity, transient adaptation, and work habits all have effects. If the viewer's acuity, sensitivity, and adaptation are poor or the eyes are older, tasks are more difficult. Even with good acuity, sensitivity and adaptation level affect momentary sensitivity.

Eyes change to maintain a sharp image. The retina becomes bleached or saturated to accomplish the best photochemical reception of the average luminance in view—pale or dark. Thus, if the field of view is mostly dark, the eyes adjust to dark. Then as the eyes move, the iris accommodates to other variations by opening or closing. As a result, if opening and closing is required over and over again, the eyes strain and become fatigued. At the same time, viewers alter the positions of their head, neck, and shoulders to maintain a sharp image, particularly if a glare is on the viewing material. As a result, if adaptation is required over and over, muscles tense and fatigue. These strains and fatigue can reduce productivity.

Likewise, the viewer's work habits affect the light requirements. Is the work long or short? (Long needs more illumination.) The direction the viewer faces also affects the light delivery. Viewers typically look in several directions—straight ahead or to the side, in a head-up or a head-down position. Thus, the light is received from a different angle. The best possible light for one position may be the worst for another.

The Surroundings

Surroundings affect the amount of light. The furniture, the room surface reflectance compared to task illuminance, and the evenness of room surface brightness have their effects.

Rules of Thumb for Interior Finishes
- Never put glass on top of a desk!
- Never have a glossy finish on the desktop.
- Always use a pale colored desk if paper-based tasks are perpetual.
- Similarly, make walls pale colored with a matte finish if desks are countertops or placed next to a wall.
- Always minimize contrast behind computer screens
- Constantly, wall, floors, cabinets, or other large surfaces in view will reflect or not, depending upon the value of their color. Design light accordingly.

The furniture can create lighting problems. For instance, overhanging upper cabinets can throw a shadow on the desktop. Surface finishes—matte or glossy—affect reflectance. Never have a completely glossy desktop; the glare is unrelenting. A desk with a computer screen hung below a desktop window needs careful ceiling fixture placement.

REFLECTANCE COMPARED TO TASK

Surfaces near and far within the cone of vision affect the visibility of the person sitting at the desk. In a heads-up position, the cone of vision is 55 degrees up from vertical. But heads are not always up. Within the cone, excessive contrasting surfaces can require the eye to adjust and readjust, resulting in fatigue. A desk used for paper-based tasks should be pale in color, creating a noncontrasting background to the task (white paper). Use pale matte surfaces for near surfaces (within arm's reach).

Likewise, near and far surfaces within the screen's cone of vision affect visibility. Excessive contrasts and glare from surfaces can inhibit seeing the screen. Screens positioned perpendicular to the floor have a cone of vision 65 to 75 degrees up from vertical. But laptop and notebook computers with different screen angles have different cones of vision. Determine the screen's cone of vision.

Cones of vision: screen; person.

Nearby surfaces are within arm's reach. For paper-based tasks, keep the ratio not more than the task footcandles and not less than $1/3$ of the task. For screen-based tasks, keep the ratio of nearby surfaces within 3:1 of the screen image to the surface behind the screen.

Rules of Thumb for Contrast Ratios of Nearby Surfaces

For paper-based tasks:

Not greater than the task illuminance nor $1/3$ less than task.

For screen-based tasks:

Not greater than 3:1 for screen image to surface behind screen.

Determine if a surface would be too bright by using point method calculations. If the amount reflected from uplight is too high, glare will fall on paper-based and screen-based tasks (particularly dark screens). In addition, uplight with uneven brightness and hot spots (greater than 10:1 contrast ratio) can adversely affect visibility.

Distant surfaces are beyond arm's reach. Glass office partitions permit distant surfaces to be seen. Not good if contrasting. Anticipate the view for each desk. Remember that distant surfaces in view may actually be in an adjacent space, like a bright sunny window in another room.

Point Method Calculations

$$\text{straight down fc} = \frac{\text{candlepower}}{D^2}$$

$$\text{horizontal surface fc} = \frac{\text{candlepower} \times \text{cosine of angle}}{D^2}$$

$$\text{vertical surface fc} = \frac{\text{candlepower} \times \text{sine of angle}}{D^2}$$

Thus, the design responsibility sometimes goes beyond the office space in order to achieve the best possible lighting.

Dark walls, floors, cabinets, or other large dark surfaces will reflect only a small amount of the light received. Such surfaces could create a harsh contrast to a brightly lighted task. In such circumstances, increase the amount of general illumination to counteract the problem.

Surround a bright window with pale surfaces so that contrast of outside sunlight and inside dimness will be diminished—particularly in sunny high rises or sunny climates with few trees. Reduce sunlight if it might reflect strongly from walls or floor. Specify interior finishes carefully to maintain reasonable contrast ratios. For screen-based tasks, keep the ratio of distant background surfaces within 10:1 screen image to background. For paper-based tasks, keep the ratio not more than 5 times the task footcandles and not less than $1/5$ of the task.

Rules of Thumb for Contrast Ratios of Distant Surfaces

For paper-based tasks:

Not greater than 5 times the task illuminance nor $1/5$ less than the task.

For screen-based tasks:

Not greater than 10:1 for screen image to distant background.

Too much contrast...

...monitor and copy equal.

EVENNESS OF SURFACE BRIGHTNESS

Evenness is controlled by luminaire engineering, placement, and distribution. On vertical surfaces, create a smooth wash of light within the acceptable ratios, both side to side and top to bottom. Ceiling surfaces should also appear even.

Rules of Thumb for Even Brightness

Vertical Surfaces
        side to side 1:1.5
        top to bottom 1:5

Ceiling Surfaces
        not more than 10:1
        best at 4:1

Determining the Amount

Sometimes, the amount of light needed is as high as 100 footcandles, but typically ranges from 20 to 75 footcandles. Determine the amount of illuminance by utilizing the Illuminating Engineering Society of North America's Recommended Illuminance Weighted Chart. This system is based on viewer's age, speed or accuracy required, reflectance of task background (paper or video screen), and length of time tasks are done. Essentially, it suggests that if the visual task is difficult and the viewer is older, use the highest amount of illuminance, and if the task is simple and the viewer is young, use the lowest amount.

Determining Light Distribution

Once the illuminance has been determined, choose the light distribution—uniform or nonuniform. On the one hand in large and small commercial offices, uniform lighting has frequently been used, not because it is better, but probably because the desk arrangements were not known. In large offices, uniform lighting can waste energy and yield poor light. On the other hand in home offices, uniform lighting has rarely been utilized probably because low-end lighting technology (leftover tabletop luminaires) has been used.

Contrary to this practice, uniform lighting for small (less than 20 feet) offices is a reasonable solution. First, due to the cone of vision, the small office user does not see the ceiling within 5 feet (1.5m) in front of the desk with an 8-foot (2.4 m) ceiling. Hence, direct glare or uneven ceiling brightness might not be seen from the desk. Likewise, a vertical screen's cone of vision can see only beyond 6 feet (1.8 m) behind the desk. Many surfaces are not "in view."

Rules of Thumb for Cone of Vision in 8' Ceiling

The viewer looking straight ahead does not see any ceiling within 5 feet of the desk.

A self-illuminated screen in a perpendicular position does not see any ceiling within 6 feet of the desk.

Second, in home offices, the task equipment is usually located in the office. By contrast, in commercial offices, several rooms contain tasks—mail room, copy room, FAX room, and video conference room. Thus, in home offices, lighting uniformly wall to wall is more defensible.

UNIFORM

Design office lighting by deciding whether the whole space must be uniformly lighted or not. Uniform lighting distributes relatively the same amount of light from wall to wall. Only use it when task surfaces fill the room and a minimal amount of light falls on the floor. Uniform lighting can be determined with zonal cavity calculations (by hand or computer) or for a very rough estimate by a weighted rule of thumb.

Zonal Cavity Method for Determining Amount of Fluorescent Downlighting

1. Determine room cavity ratio.

$$\text{cavity ratio} = \frac{5 \times \text{room cavity} \times (\text{length} + \text{width})}{\text{length} \times \text{width}}$$

2. Look up coefficient of utilization using room cavity ratio.

3. Calculate # fixtures with zonal cavity formula.

$$\text{\# fixtures} = \frac{\text{fc desired} \times \text{total room area}}{\text{\#sources/fixture} \times \text{lumens/source} \times \text{coefficient of utilization} \times \text{light loss}}$$

Rule of Thumb for Determining Fluorescent Downlighting

$$\text{\# sources} = \frac{3 \times \text{room area of the space} \times \text{fc desired}}{\text{lumens per source}}$$

General Electric.

But uplight might be more desirable. If so, the ceiling needs to be matte white. Uplight works in spaces with 9- to 12-ft high (2.7 to 3.6 m) ceilings. Too low a ceiling creates hot spots and too high reduces reflected light. Neither provides the best possible light.

Uplight only is bland! Add some direct light —a ceiling-mounted accent light, a portable luminaire, or a window. Uplight must conform to specific criteria to be the best. The criteria are even illuminance, amount of luminance on the work surface, and brightness ratios. Brightness from uplight is more tolerable than brightness from direct light, because direct-light luminaires appear as bright patches against a dimly lit ceiling. To calculate uplight, use either of the fluorescent downlighting formulae and add 30 percent more.

Rule of Thumb for Fluorescent Uplighting
sources = downlight # sources x 1.3

NONUNIFORM

Uniform lighting is not always the best. Nonuniform lighting makes some areas brighter than others, determined by task requirements and desired effects. Put high footcandles at the desk and spread low footcandles elsewhere. Nonuniform can be more energy efficient and more interesting. Nonuniform lighting can provide accent lighting to counteract the cave-like effect of the louver distribution in commercial offices. Nonuniform lighting can be tailored to the home office user with great creativity, since the home office user is also the final lighting decision maker. By contrast, in commercial offices, the corporate or governmental budget and the upstairs management are the decision makers.

THE BEST DISTRIBUTION

The best distribution for nonuniform lighting is both direct and indirect, with indirect being at least 25 percent of the total. Combining lighting avoids the dull and bland appearance of totally indirect lighting and illuminates the task first and the space second. Not a bad scheme!

Rule of Thumb for Best Possible Lighting Distribution
Direct + at least 25% Indirect

SOURCES FOR DIRECT OR INDIRECT

A large area of direct light can be produced by fluorescent with 8- to 12-foot (2.4 to 3.7 m) ceilings and by HID sources with higher ceilings. Indirect light can be produced by linear sources (fluorescent) or by point sources (metal-halide, white sodium, or high-wattage incandescent). Metal-halide costs more to purchase and install than fluorescent, but uses less electricity. Therefore, the higher initial cost is offset by lower operating costs. Incandescent is the least costly to purchase, but the most costly to operate.

A small area of direct light (accent light) can be produced by line- or low-voltage incandescent, HID, or compact fluorescent sources. HID's work with high ceilings. Incandescent work with lower ceilings, particularly low voltage in a low wattage. Compact fluorescent sources with well-designed reflectors can also deliver accent light— be careful of the color of the light.

LUMINAIRES FOR DIRECT AND INDIRECT

Luminaires for direct light can be ceiling re-

Small-cell parabolic louvers can create a gloomy scene...

...unless upper-wall brightness is provided.

cessed, surface-mounted, or suspended. If installed with flexible wiring in suspended-grid ceilings, they can change when the desk changes.

Luminaires for uplight put light on the ceiling—a large architectural surface. They can be suspended, wall-mounted, furniture-integrated, or freestanding. Suspended uplights without brightness strips are seen as dark images in front of a bright ceiling as opposed to recessed luminaires which are seen as bright patches against a dark ceiling. Thus, plan the visual pattern the luminaires will make; it is ultimately seen!

Furniture-integrated or freestanding luminaires have several financial benefits. They are classified as furniture for tax purposes. Therefore, they can usually be deducted at the accelerated rate of furniture rather than at the slower rate as building equipment. They do not disturb the fire-rated ceiling of a structure. Thus, a lower fire insurance rate is possible.

Many considerations determine which system is the best, particularly for commercial offices. They include aesthetics, cost of installation, tax break, insurance rate, cost of maintenance, cost of electricity over time, and available funds. Compare, compare, compare.

Amount Can Hinder

In addition to being helpful, light can hinder work if it is too little or too much and if it creates glare or excessive contrast. If any one of these conditions exists, the visual task is made more difficult and the eyes and body respond.

If the long-term reading surface (either screen or paper) is too bright and the surroundings too dull, the eyes exert to adjust to both conditions over and over. Therefore, teach your clients always to have two lights on in a room where they are doing any long-term work.

Light can also hinder when it reflects from a surface that is too bright or unevenly lighted. Therefore, keep surface within view of a desk evenly lighted and within the acceptable contrast ratios.

Light can hinder even if it is the correct amount but comes from an incorrect position. Thus, position of the luminaire is critical for the best possible light.

the best color

Each source has its own color of light. For paper-based tasks, the color should maximize the contrast of black print on white paper. Choose a source with as much blue-green as possible (around 510 nanometers). According to research at the Lawrence Berkeley Labs, blue-green assists the eye's adaptation process and enhances reading of paper-based tasks. Therefore, when using fluorescent sources, examine the spectral distribution curves and select the one with the most blue-green. Typically, high Kelvin sources have more blue-green; low Kelvin sources have less. Thus, for small offices, discourage clients from using incandescent and 28K compact fluorescent sources.

True colors (of tasks and surroundings) can be distorted or revealed depending upon the illuminance level and source type. Sometimes seeing true task colors might be critical. The range of IESNA's recommended illuminance for desk tasks is 20 to 100 footcandles. Within this range, incandescent sources providing more than 50 footcandles of light can make colors appear unnatural. Within this range, fluorescent sources above 30K will make colors appear natural. Thus, 28K compact sources are not the best possible choice.

In addition, be aware that a dominant surrounding color can also affect the impression of "making colors appear natural." For example, 41K fluorescent sources in a totally blue-gray room (carpet, walls, shutters, upholstery) will appear unpleasantly icy-blue. The decor needs to be warmed up with warmer light sources to counteract the blue-gray environmental impression. Color is tricky; use it wisely.

the best position

Various positions deliver glare. Even though the amount is sufficient, any glare can reduce visibility. Glare can be direct from the luminaire or can be indirectly reflected from a surface. In addition, the best position is also controlled by the habits of the viewer and where the various tasks are done. For example, the best possible desk light facing forward becomes the worst possible when the viewer turns left or right to read. Position is everything in light delivery!

Direct glare is rated by VCP.   Reprinted by permission of Lightolier.

Direct Glare

In large commercial offices (more than 20 feet), direct glare can come from rows of ceiling luminaires. The eyes must constantly adapt to bright light above and dimmer below, as the office worker looks up, down, up, and down. Therefore, choose ceiling luminaires for visual comfort.

Use the Visual Comfort Probability (VCP) rating for choosing direct-light luminaires. VCP predicts visual comfort based on people viewing the luminaires from the worst possible position in the space—the back of the room looking at a ceiling full of luminaires. VCP ratings are listed in luminaire manufacturers' technical data. A VCP rating of 80 or more is good. Such a rating will assist in getting the best possible light.

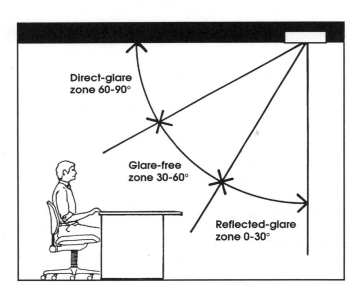

For screen-based tasks, the luminaire should not produce light above the 60-degree angle (with 0 as straight down) within the person's view. Some uplight luminaires have a brightness strip or

mesh along the sides. Make sure their direct light is not glaring.

Be aware that fluorescent luminaires with different louver-cell configurations deliver different qualities of light from different directions:

- 2 x 4 luminaires can produce veiling reflections from the 4 ft sides.

- 2 x 2 9-cell fluorescent luminaires deliver equally from both sides.

- 2 x 2 16-cell fluorescent luminaires deliver more light directly below and less to the sides.

Reflected Glare

If direct light comes from less than a 30-degree angle to the task (with 0 as straight down), glare reflects from any glossy reading material—photographs, glossy paper, and computer screens. The reflected glare produces a blur, destroying the readability and requiring the person's head or the glossy surface to move. The blur is called a veiling reflection. In the long run, discomfort and loss of concentration can occur, reducing productivity.

See for Yourself: Is There Reflected Glare at the Desk?
1. Put a small pocket mirror on the desktop where the paper-based task will be.
2. Can a luminaire be seen in the mirror?
3. If so, reflected glare will be received.

For paper-based tasks, position the luminaire to reflect light away from the eyes. (Light leaves a surface at the same angle but in the opposite direction from which it arrives.) Consequently, luminaire positions above and in front of paper-based tasks create veiling reflections. Positions behind or at the sides do not.

Light should reflect away from not into your eyes.

Solutions for Glare

If glare will be received, change the position of the luminaire, the desk, or the task location. If the luminaire causing the glare cannot be moved, try a polarizing lens, or move the desk to get the best possible light from another position:

- Rearrange the desk with two rows of ceiling fluorescent luminaires on either side. The light will come from both the right and the left, balanced and shadow-free.

- Turn the desk around to face fewer rows of luminaires.

- Move the desk so the luminaire is directly over the person's head.

screen-based tasks

Screen-based tasks require well-planned solutions. The surroundings profoundly affect visibility. Utilize the *Rules of Thumb* for surroundings in this chapter. First, the surroundings should not distract the eye from the task. Surfaces should be even in brightness. For instance, a window in view should be covered in a material that has the same color value as the wall.

Second, provide light without glare. The screen should not face a lighted source or a bright surface. Otherwise, the reflection could obscure the information displayed. Rooms with multiple computers are a challenge—such as stock exchanges, traffic control rooms, and security offices. If lit with evenly bright uplight, a uniform veil would be on all the screens with no spots of glare and could be the best possible distribution.

Third, the paper-based copy for computer input should not be exceedingly brighter than the computer screen. The older the worker, the more illuminance needed. The poorer the contrast between print and paper, the higher the amount of illuminance required, unless the screen is dark. If so, change the screen's color to white.

Fourth, provide light for both tasks. Paper-based tasks require light that comes to the paper and is reflected away from the eyes (usually fixtures behind the person) and screen-based tasks require not lighting the screen (usually fixtures in front, but not far enough to get into the 60- to 90-degree light-delivery zone). Consequently, if possible, separately switch the ceiling luminaire for each task—one switch for paper-based; the second for screen-based. Dimmers (wall-mounted or remote) can control these changes.

Further, if using undercabinet or shelf-hung luminaires, use batwing or bilateral lenses. If using desktop luminaires, use adjustable ones to position without lighting the screen.

Fifth, assure that the visual effect of the space is pleasing. For instance, a space lighted only with parabolic-louvered luminaires may appear oppressive unless relieved by some vertical-surface light.

Sixth, the luminance of computer screens ranges from 5 to 25, with 15 for the average. The luminance can be changed to fit the luminance in the space.

For screen-based tasks.

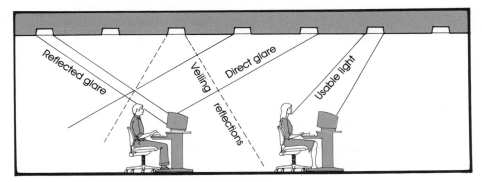

Paper-based tasks occur anywhere—the desktop, the lap, the FAX machine. Thus, the designer must provide for many task situations. The visibility varies widely due to many inherent characteristics. The best possible paper-based task has a high contrast of image (letters) to the paper. Matte white paper typically has a reflectance value of 75 percent—not really high. Overall, the best possible light is high enough in amount for the task and the user, is delivered from a nonglare angle, is in a color that enhances vision, and is adjustable in amount of light and/or position.

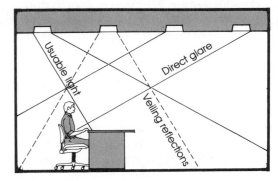

For paper-based tasks.

Portable Luminaires

A floor, table, or wall-mounted portable luminaire used for desk light must deliver light well over the desktop with no glare. Position a single desk luminaire to avoid shadow—opposite the hand used for writing. Whenever possible, purchase a luminaire with two levels of light—high and low. Likewise, purchase an adjustable luminaire, giving greater range of positions. Use fluorescent; they are cool, energy efficient and obtainable in 41K to enhance vision. If choosing incandescent, halogen is the highest Kelvin. Low-voltage halogen luminaires are small and can be low wattage. One manufacturer has combined metal-halide and line-voltage incandescent sources in traditional portable luminaires for higher Kelvin and brighter light.

If fluorescent:

- Position to the left or to the right, never in front of the paper-based task.
- In front of the task, get a batwing or bilateral ray lens.

- Put a pair of luminaires on both sides, balancing the light.
- Get 9-, 13- and 28-watts compact, but these might not give enough light for difficult visual tasks.
- Get a compact luminaire that accepts a 41K source.
- Specify an electronic ballast—very quiet.
- Use 41K deluxe sources for long term paper-based tasks.
- Provide 20 to 30 footcandles.

If line-voltage incandescent:

- Use a three-way source (incandescent, 2D, or circle fluorescent).
- Use a translucent shade.
- Have the bottom of the shade 16 inches wide.
- Have the bottom of the shade level with the eyes.
- Use a glass bowl diffuser under the shade.
- Use a coated source.

If low-voltage incandescent:

- Use the lowest wattage possible to keep the desktop cool.
- Allow space to hide the small transformer on the cord.
- Get several source replacements because they may be hard to find.
- Handle open halogen sources with gloves so the source will not shatter due to oil from hands.
- Choose a luminaire with a lens if using open halogen sources. Or, choose a glass-enclosed halogen.

Sample Electric Cost

A 20-watt compact fluorescent and a 100-watt incandescent source put out 1,000 lumens. The compact fluorescent when used 4 hrs/day at 10¢ per kilowatt-hour saves 96¢/month and the source lasts 13 times longer.

Combining Daylight and Electric Lighting

Balance changing daylight throughout the day with changeable electric light and window covering control.

- The window should be opposite the hand used for writing.
- Replace diminishing daylight with electric light from the same side of the room.
- Equalize too bright a window with brighter room illumination.

- Illuminate the desktop with enough electric light, if the window cannot provide it.
- If the electric light is to be 90 percent indirect, daylight relieves the bland effect of totally indirect light.
- If the window is in front of the task and could cause a veiling reflection, increase the amount of electric light.
- At certain times of the day, daylight can counteract the light from a ceiling luminaire that causes a veiling reflection.
- In offices with no windows, include some direct electric light to resemble daylight.

Put a desk luminaire to the side of the reading material.

at the office

In commercial offices, lighting has the power to create a productive work environment. Research has shown that when the light level is low, visual efficiency is poor and work can slow down. To convince management that productivity would be increased with glare-free light, call attention to the fact that the average worker's earnings, including fringe benefits, cost the company more than 200 times the electricity used by that company. Therefore, the best possible lighting can enhance productivity and save operating dollars. Further, because of its impact, lighting could improve the most but cost the least. Lighting benefits can be quantified. Light source and lighting control manufacturers can help. Overall, good light is good for business.

In states with power budgets limiting the amount of electrical consumption, designers must work with the full design team to trade off electrical use to achieve the best possible lighting.

Large Offices

The size of the office and its furniture affects lighting choices. Large multiperson offices have many lighting problems. The office size puts recessed luminaires and their lenses in view. An overly bright lens can produce glare. Louvers viewed at certain angles can also produce glare. Use a Visual Comfort Probability (VCP) rating of 80 or more to assure comfortable lighting. Further, light distribution from small-cell parabolic louvers should be relieved by accent light in view.

Fluorescent luminaires are the best for direct light with ceilings 8- to 10-feet (2.4 to 3 m)

high. Incandescent downlights are the worst. A fluorescent luminaire produces different amounts of illuminance parallel, perpendicular, and diagonally below 90 degrees. Use rapid start 4-foot (1.2 m) T-12 or T-8, or 2-foot (.6 m) TT-5 long compacts. TT-5s come in 39, 40, or 50 watts. One-foot T-8 or T-10 and compact fluorescent (18 watts) are made for 1 X 1 grid ceilings. All can be dimmed if using a dimming ballast.

Put direct-light luminaires
over the chair...

...not over the desk.

Direct light should come from different directions for paper-based and screen-based tasks. This requirement is difficult with luminaires switched all on or all off. Hand-held remote dimmers and fluorescent dimming ballasts solve the problem, allowing individuals to control the light.

Indirect light can be produced by cove, pendant, furniture-integrated or freestanding luminaires. The sources can be high-intensity discharge (HID), fluorescent, standard or high lumen output, or remote sources with prismatic film or tubes (the most energy efficient but the most expensive initially).

PENDANT UPLIGHT

Pendant luminaires can provide uplight with fluorescent or HID sources. Some fluorescent pendants can deliver the light broadly with 8-foot (2.4 m) ceilings for a suitable luminance ratio. Some are beautifully engineered to create a luminance ratio of 4:1. Poorly engineered pendants with low ceilings can cause unpleasant ceiling-surface contrasts of 40:1 (hot spots). Pendant manufacturers have responded to the research indicating that uplight only visually appears less bright than uplight with some direct light. Thus, some pendants are designed with brightness strips or mesh the length of the luminaire to de-

liver direct light. Light source choices are T-12, T-8, or TT-5 long compact fluorescent. Color choices for some luminaires are vast and, happily, aesthetics can be part of the technical decision.

HID pendants must be hung far enough from the ceiling to reduce hot spots of light. With ceilings 10 feet (3 m) or higher, lighting distribution is easier for any pendant.

FURNITURE-MOUNTED OR FREESTANDING

Furniture-integrated or freestanding high-intensity discharge (HID) uplighting can easily be far enough away from the ceiling to allow the light to spread and blend. The sources are 70- to 400-watt metal-halide or 100-watt white sodium. Do not make the ceiling excessively bright. Luminaires could be furniture mounted on landscape-office dividers or freestanding on the floor.

When using HID indirect light, also provide some direct light in a compatible color. With metal-halide use low-voltage MR or PAR; with white sodium use 30 or 35K fluorescent. They blend the best.

LANDSCAPE-OFFICE LIGHTING

Landscape offices are large spaces with many desks hung on partitions. They present lighting problems. Ceiling-mounted luminaires spaced to yield uniform illumination often do not, because partitions and overhangs block the light. Also, uniform lighting is not completely uniform and some task locations receive less light than intended. Offices with landscape-furniture systems require undercabinet, lighted shelf, and/or portable luminaires.

Landscape desks can be moved, usually needing, but not getting, ceiling luminaires moved. Lighting can change when desks change if lighting is movable:

- Ceiling-mounted direct luminaires with flexible wiring in a suspended-grid ceiling.
- Direct-light luminaires on uppercabinet, shelf, or desktop portable luminaires.
- Furniture-mounted or freestanding uplight luminaires.

Undercabinet Luminaires

Undercabinet luminaires for desks need to be engineered differently than undercabinet luminaires for kitchens. Reflected glare must be eliminated so that reading tasks will be visible. Manu-

Office landscape-furniture systems require task lighting.

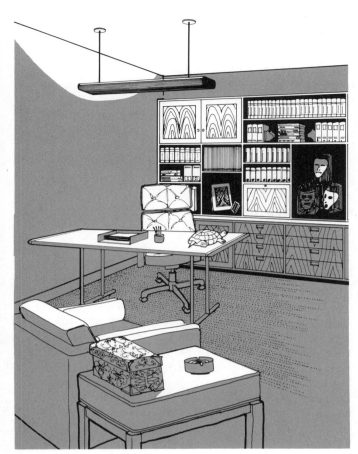

Often uplight can supply better light than direct light.

facturers have carefully studied the reflected-glare problem and have developed several solutions: lenses, reflectors, louvers, and movable luminaires.

Batwing or bilateral lenses, some louvers, and some reflectors throw the light left and right. Two special fluorescent track-mounted luminaires can move in or out for greater or lesser amounts of glare-free light. All developments have been innovative. All attack the same problem. End results are dissimilar. Choose the one that works for the installation.

Batwing lenses work best when the reader is facing directly ahead. Facing at an angle can result in veiling reflections. A diffused batwing spreads the light's intensity and is a better choice.

Often furniture manufacturers include luminaires with the desk which ignore the glare problems. Undercabinet luminaires can be evaluated by luminance-contrast meters that map the work surface for lighted area and contrast reduction percentages (loss of visibility). Choose the luminaire with the largest area light and the least contrast reduction. Diffused batwing lenses win.

Shelf Luminaires

Shelves with undercabinet luminaires below are excellent for a desk if they do not create reflected glare. Follow the guidelines for undercabinet luminaires. Install the shelf 15 to 18 inches (38 to 46 cm) above the desktop and 9 to 12 inches (23 to 30 cm) back from the front edge of the desk.

Furniture-Mounted or Freestanding Luminaires

Furniture-mounted or freestanding luminaires should not destroy the uniform brightness of the ceiling when moved. Care must be taken to have the best possible lighting in a multiperson office after the move. Encourage clients to install flexible wiring for their ceiling-mounted, direct-light luminaires. Then they could be moved also.

Calculations for Landscape Offices

Calculations and design must be done carefully for landscape-office lighting. Watts per square foot or square meter do not provide visibility. Spacing criteria do not work if partitions and overhangs are present. Also, surface reflectance values are not constant because some surfaces are in shadow, diminishing the reflectance. Consequently, calculations based on coefficient of utilization are often inaccurate. Such calculations usually indicate that the task will receive more light than it does. Steven Orfield (see Bibliography) suggests that the most important variables for landscape-office lighting are:

- distance from light source to the task
- orientation of the source
- intensity of source in task direction.

Further, he suggests that solutions be defined in these terms before products are considered. He calls this method "coefficient of application." Use it rather than the coefficient of utilization.

Small Offices

Small offices (less than 20 feet) have less lighting problems than large offices. In small offices, the ceiling luminaires are sometimes not in view. Consequently, direct glare from ceiling luminaires and evenness of indirect light on ceilings is not as apparent. However, the rules hold true for avoiding veiling reflections, avoiding window glare, and needing some direct light to counter-act mostly indirect light.

In a small office, it is tempting to position a recessed luminaire where it architecturally seems correct—over the desk. Do not do it! That position is the veiling reflection zone and glare will occur on paper-based tasks. Also, do not put it too far behind the person at a computer. That position is the reflected-glare zone, and glare will occur on the screen-based task. The solutions are to put one luminaire in front of and one behind the desk and separately switch, to use one luminaire on each side of the desk, or to use indirect light. Cleverness can overcome the glare-zone problems.

Luminaires for uplight could be uplight brackets (wall mounted) or cove light (wall or furniture mounted), as long as the brightness at the top of the wall and the ceiling are not in the screen's cone of vision. Uplight works well for small offices, particularly for traditional-style offices where sources should not be seen. Use fluorescent. Install the uplight luminaire at least 10 inches (25.4 cm) down

Position one ceiling fixture over the chair, not the desk.

from the ceiling. The farther down, the greater the spread of light. Choose the height of the faceboard as required to hide the light source from view when seated and standing. Reflectors or angled blocking behind fluorescent sources can aim light toward the ceiling. But reflectors also collect dust reducing the light. Request that the coves get dusted regularly.

Remote source lighting with hollow light guides for coves can transfer light evenly from a single MR-16 (or projector HID) source. The system is energy efficient and not heat producing. The brightness is determined by the source type, wattage, and lumens.

Movable Files and Fixed-Library Shelves

Lighting movable files is tricky. Files are not always where they were or where they should be to receive the best light. Likewise, the end wall containing access information must be highly visible. For the most part, reading at the files will probably be for identification and not for comprehension. Consequently, the best solution is to create a suitable amount of ambient light. If the filing system is color coded, consider deluxe 41K fluorescent that renders colors well.

Lighting fixed-library shelves is easier. They do not move. Direct-light luminaires can be attached to the top of the shelves with linear sources and reflectors to spread down the face of the shelves. The farther out from the shelf, the farther down the light will spread. Attach luminaires at least 8 inches (20 cm) out from the shelves. The luminaire should block the view of the source from the common positions in the room, both seated and standing. A lens or louver can cover the opening, if desired. Source choices can be linear fluorescent, linear incandescent, or remote-source lighting with custom-made cornice and prismatic film to transfer light a long way.

Asymmetric-fluorescent, ceiling-mounted luminaires provide library shelf light. Also, recessed, perimeter wall luminaires work well when the shelves are also the wall.

Lighting movable files is tricky.

at home

At home, all the principles for office lighting apply. But the home has different constraints and luminaire acceptability. First, the rules for paper-based and screen-based tasks should be followed. Poor visibility and eye fatigue is not limited to the commercial office.

Second, short-term tasks are done in many rooms not intended for office work. Thus, accommodations to residential luminaires must be made. The luminaires in the room might not produce the best possible lighting for office work, but with planning, they can be suitable. For example, an eating table sometimes becomes a short-term desk. A chandelier over the eating table creates general illumination usually too dim for paper-based tasks. Choose a chandelier with a downlight separately switched. Unfortunately, such light will produce veiling reflections on glossy tasks, but the downlight will be the difference between seeing or not seeing other paper-based tasks. With dark screen-based tasks, have your client turn off the downlight, eliminating too bright a background,

which could cause eye fatigue.

Most any household luminaires—floor, table, or wall—can be used as desk lights if they meet these five requirements:

- A shade that transmits light.
- The bottom of the shade 16 in. (41 cm) wide.
- The bottom of the shade level with the eyes.
- A glass bowl diffuser under the shade, if an incandescent source.
- The ability to provide over 1,000 lumens.

Third, home offices are becoming the skyscrapers of yesterday. This change is due to corporate downsizing and decentralizing, supported by electronic transmission of data. More and more of these home offices are in dedicated spaces. More and more have screen-based tasks screens (home automation touch screens, video phones, and satellite TV). Highly engineered uplight luminaires for commercial spaces can be used in dedicated home offices effectively.

Many screen-based tasks are in offices.

mementos or art. Neon sculpture could be used, if not distracting. Stunning general lighting could come from well-placed, highly-engineered commercial uplights in one of the luscious colors offered by some manufacturers. Or remote-source lighting can carry light around the room in coves.

Well-designed luminaires for desks could be elegant, tall, table-top luminaires with two levels of illuminance—high for prolonged reading and low for notebook computer input. One manufacturer has combined metal-halide with incandescent for higher illuminance. Others offer a less energy-efficient 3-way switch for incandescent. Further, long-arm, adjustable luminaires can accommodate two adjacent work surfaces from wall, floor, or tabletop positions. Personalization is only limited by the pocketbook, not the imagination.

Seventh, desks in kitchens with upper cabinets require undercabinet luminaires that redirect the light into the glare-free positions for paper-based tasks. Do not specify kitchen undercabinet luminaires. They create glare!

Eighth, desks against the wall could have lighted shelves above.

Ninth, a surface-mounted, fluorescent ceiling luminaire positioned over the person's head

Light from both sides avoids shadows.

Fourth, small, drop-lid, or roll-top desks have no space for a desktop luminaire. General illumination can provide short-term desk light when no other solution works. Well-spread general illumination can come from a floor-based uplight with a dimmer, as long as it is not the only light on in the room. Or, general illumination can come from pencil-thin fluorescent strips behind elaborate cove molding near the ceiling.

Fifth, a linear, double pendant (library light) or movable pendant can accommodate two desk users. For example, a linear, double pendant can illuminate from both sides with glare-free light and be separately switched if two people use the same desk but write with different hands. Likewise, the same effect can be obtained from a pendant adapted to a ceiling track and moved from left to right as required. A single pendant should hang on the opposite side of the hand used for writing.

Sixth, lighting home offices can be highly personalized since the decision maker is the office user. Thus, creative accent lighting, stunning general lighting, as well as unique desktop luminaires can be utilized. Creative accent lighting could be cable luminaires accenting personal

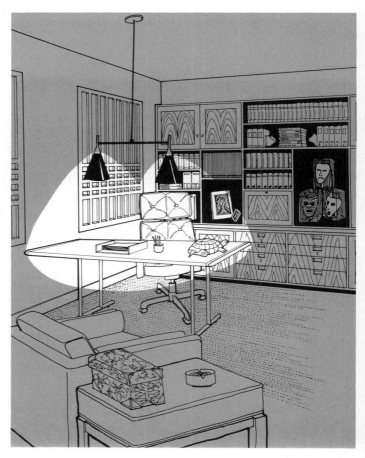

provides work light. Never have a person's back to the light if it creates shadows. Avoid all overhead, open incandescent downlights—the light is harsh.

Tenth, teach your client that the light at the desk should not be the only light on. Create high-quality desk lighting at home—it is truly an office!

energy conservation

Energy can be conserved in offices beyond what governmental agencies require (or do not require). Lights can be turned off when not in use. Energy-efficient sources and well-engineered luminaires can deliver the best possible light. Highly efficient louvers (67 as opposed to 41 percent) can deliver the most light. Electronic ballasts conserve energy better than magnetic ballasts. Reflectors designed specifically for the source will deliver more light for the watts consumed. Cool sources (fluorescent as opposed to incandescent) will minimize heat intrusion.

Sample Electric Cost

Retrofitting hall and stair wall-mounted luminaires with 9W compact fluorescent sources instead of 60W *A* incandescent saves $20.40 per year at 10¢ per kilowatt-hour and used every day.

$$\frac{51\text{W saved} \times 4{,}000 \text{ hrs/year} \times 10¢/\text{kilo. hr}}{1{,}000} = \$20.40$$

In some offices, the light level can be reduced. Determine how sensitive the tasks will be to the level of illuminance.

In spaces where lower levels are acceptable:

- Replace every other source in a rapid-start luminaire with an energy-saving source and reduce energy consumption by 33 to 50 percent. All sources will appear equally bright.

- Install power reducers in-line with the ballast.

- Reduce wattage to 34, 32, or 28 for 4-ft (1.2 m) luminaires.

- Specify reflectors.

- Retrofit polarizing lenses in luminaires that now have prismatic lenses and remove two of the four sources per luminaire.

- Use utility company rebates for energy-conservation, such as lens replacement, source removal, and high-power ballasts.

In offices, the benefits of reducing energy consumption should not be offset by loss of productivity as long as the light for tasks comes from the best angle for visibility and yields the amount needed for the age of the worker, the contrast of the task, and the speed and accuracy required.

In spaces where higher levels are needed:

- Use sources with higher efficacy (number of lumens produced for watts consumed).

- Specify rare-earth or triphosphor fluorescent.

- Utilize energy-saving electronic ballasts, as opposed to magnetic ballasts and save 8 or more watts for each two-source, 40-watt luminaire.

- Replace lenses in fluorescent luminaires that put light in the glare-producing zone (60° to 90° above vertical) with lenses that put light into the 30° to 60° zone (batwing or polarizing). Task visibility can be enhanced.

Automatic Lighting Controls

Office work has people coming in and going out. Frequently, lights are left on unnecessarily. Controls can automatically switch lights as needed, thanks to microprocessor technology. Further, controls can dim lights when daylight is available and reduce energy by 20 to 50 percent. Controls are activated by time clocks or sensors. Sensors respond either to light or to people (their motion or their heat).

Time clocks are reliable for turning off lights. They can automatically adjust to seasonal time changes, can be manually controlled, and can be staggered when turning on after electrical failure to prevent damaging power surges. Likewise, they can limit manual overrides.

Photosensors respond to light. They can be preset for a specific amount of light and respond by dimming or by turning on/off. Likewise, photosensors can control several areas at preset levels—such as an illuminated sign at 100 footcandles, drafting office at 200, and computer office with dark screens at 20.

Ultrasonic sensors are people-sensing controls responding to motion. They are triggered (hopefully) by the right motion and the best ones remain triggered even though motion momentarily ceases during occupancy. Otherwise, the lights turn off, as they did at a middle school in Tallahassee. The students loved the disruption; the teacher did not find it amusing.

Infrared sensors are people-sensing controls responding to heat. They are less likely than ultrasonic sensors to read inappropriate clues. But the sensor must be carefully positioned and calibrated to operate correctly. Infrared sensors work well for spaces up to about 800 sq ft (74 sq m). They are excellent security devices.

Controls should be selected to conform to the use and type of space. A people-sensing control does well with intermittent use (not a middle school). A time clock does well with scheduled use (a school or restaurant). A photosensor responding to daylight does well with a large space with windows along one or more walls (a high rise, multiperson office).

Several controls can save energy for small businesses. Large businesses need a more complex control system in a centralized location. All controls can have manual overrides for personal influence over lighting.

Some controls receive signals through the electrical lines. Do not use electric lines for signals with high-intensity discharge (HID) sources, because the HID ballast might short out the control. Controls are delicate devices!

Payback time varies. It can be as little as nine months or as long as several years. It depends upon the size of the business, the length of time used, and the local electrical costs. Once paid back, payback becomes welcomed savings.

installation

The type of ceiling construction determines the options for luminaire installation. Construction can be wood-ceiling joists, suspended grids, open-bar joists, or concrete and steel. Wood-ceiling joists, suspended grids and bar joists accept recessed or surface mounted luminaires. If the ceiling is concrete and steel, the luminaires can be surface mounted if electrical access is possible, unless they are set in before the concrete is poured.

Ceilings can be designed to integrate the lighting, such as a large oval cove with remote source lighting over a large oval conference table or a tray ceiling with perimeter downlights outlining the shape and visually expanding the small square office. Do not neglect the opportunity to develop a unique architectural ceiling to support the lighting.

The decision of which luminaires to buy depends on many decisions and trade-offs: ability to provide light needed, initial cost, operating cost, ease of maintenance, payback time, and aesthetics. Choose wisely; well-engineered luminaires and energy-efficient sources pay back over time.

Overall, understand how interior finishes and distribution of light affects visibility; how glare, both direct and reflected, can be controlled; and the requirements of paper- and screen-based tasks. The best possible desk light can be achieved.

smart
lighting for
businesses
20

Small businesses often occupy spaces of 300 to 3,000 sq ft (28 to 279 sq m) in large cities and small towns. Some sell products such as clothes, food, jewelry, flowers, or picture frames. Some provide services such as insurance, real estate, medical, haircutting, travel, or banking. All are competitive. All want to create a positive impression. Competition heightens the need to have a positive impression and also the need to be productive. Impressions are projected to a great degree by lighting. Productivity is affected to a great degree by lighting. Lighting research in offices and factories has shown that a correlation exists between light levels and productivity. It was found that, within limits, when the light level goes down, so does productivity; when it rises, productivity rises. Other research has shown that employees' morale, motivation, and safety are also influenced by how much light is available. Further, research in retail stores has shown that sales increase between 7 to 15 percent when products are accented by light. Lighting can have power.

power for selling

Light is powerful for selling products. It draws customers' attention—the first step to buying. Other senses assist, but without a doubt light makes possible the visual part of the buying decision, and visual appeal is responsible for most impulse sales. Visual appeal induces the customer to want the product. Well-designed store lighting guarantees visibility. It can make products vivid and focus customers' attention. It becomes the silent persuader. Needless to say, if the product is large, visibility is easier. If the color of the product is similar to its background, visibility is hindered and more light is required. If the color of the product is dark, details are less visible. Light permits identification of a product and reveals its parts. Products are moved frequently; lighting needs to move also.

Products differ, and different kinds of stores require different lighting. For example, clothing stores need soft general light, brilliant displays, and sufficient light focused on important areas such as cashier stations. Conversely, food and drug stores need bright general light, well spread and delivered from direct sources. This type of light reveals colors of products, creates sparkle, and commands a brisk atmosphere. The light sources should be hidden as much as possible in all stores by high ceilings, baffles, or ceiling systems. Recessed trapezoidal ceiling systems obscure lighting fixtures well by

Light a display three times as bright
as its surroundings. Adapted with permission from p. 75
of *Contract* Magazine, December 1979.

creating a visual barrier from the usual viewing angle (45°).

Rules of Thumb for Commanding Customers' Attention

- Light a display at least three times as bright as its surroundings.
- Sometimes special displays need to be five to ten times as bright.
- Products should visually dominate a commercial space.
- Illuminate display and appraisal areas at similar levels. (Exceedingly different levels complicate buying decisions. Therefore, one area should not be more than three times brighter than another. For example, if sweaters are displayed in a showcase and appraised in a dressing room, the showcase should not be more than three times as bright as the dressing room.)
- Skin tones need to be enhanced, not destroyed, by the light at mirrors. (Customers need to feel good about what they see in the mirror. Consult the details on mirrors in Chapter 9.)
- Illuminate products with the same color of light under which they will be used. (This rule applies not only to clothes but also to household products. For example, in a frame-it-yourself shop, customers coordinate their artwork with the choices of frames and mat colors. These products are chosen in the shop and must look the same when they are hung on the wall at home or in the office. Therefore, light sources in the shop should permit judging how the colors would appear under incandescent light for residences, cool-white fluorescent for offices, and

maybe sodium and metal-halide for other commercial spaces.)

Rule of Thumb for Contrast of Visible Areas of Light within a Space

Visible areas should not contrast severely with one another—no more than one to five times brighter. (At any one time an employee should not be able to see a lighted area that is more than five times brighter than the dullest area.)

Product Displays

Where are products displayed? Some displays are horizontal; some are vertical. Horizontal displays are on table or countertops, or in showcases. Vertical displays are in racks, wall hung, or freestanding. (See Chapter 11 to learn how different materials are enhanced by light from different locations.)

HORIZONTAL DISPLAYS

Horizontal displays require downlighting, showcase, or countertop lighting. Downlighting is produced by overhead fixtures. If the display is in a glass showcase, minimize glare by positioning the fixtures above the front edge of the case. Avoid downlights at mirrors, however; light directly above customer's heads makes unattractive shadows and distorts facial features. Downlight is produced by incandescent R (reflector) and PAR sources in 50 to 300 watts, or by the white-sodium T-10 sources in 35 to 100 watts. The sodium sources produce an excellent color of light while rendering colors of objects well. However, they do take up to 4 minutes to warm up to on and require 1/2 minute to restart after being turned off. These minor inconveniences can be overcome in the total lighting design with additional sources.

Showcase lighting can be fluorescent or low-voltage incandescent tubes or tapes (1/4 to 5 watts). Showcases are enclosed and heat builds up. Fluorescent is the coolest. One low-voltage showcase system has a fan for heat removal. Attract attention to the case by using three times as much light within it as around it. (See Chapter 11 for ways to light furniture.)

Countertop lighting can be direct or diffused. Direct light is intense and focused on a product. It is produced by regular- or low-voltage incandescent sources. Regular voltage requires reflector spots in 25 watts or more, depending on surrounding light. Low voltage requires a fixture with a transformer, like a high-intensity lamp. Either way, the fixtures should change. (See Chapter 4.) On the other hand, diffused is soft and spreads light in the area of products. It can be produced by table lamps with compact fluorescent screw-base of 5 to 20 watts for cool light, equal to 25 to 75 incandescent watts of light. Table lamps do not need to be bright; use them to give a sense of scale.

VERTICAL DISPLAYS

Vertical displays require accent, valance, cornice, luminous panel, or shelf-edge lighting. Accent lighting is confined and intense. It can come from incandescent or high-intensity discharge (HID) sources. On the other hand, light from valances, cornices, and luminous panels can be well spread and less intense.

Accent

Accent light is used for feature displays. Normally, it is created by a point source. It should reveal details and attract attention but not detract from the rest of the space.

Rule of Thumb for Accent Light
The center of the light should fall on the important parts of the display at a 60% angle.

---

60 degree angle.

This rule ensures the most effective lighting with natural-looking highlights and shadows, and it minimizes glare for customers approaching from the other side.

Accent fixtures should be adjustable. Use recessed, surface-mounted, or track fixtures. Tracks accept additional fixtures for more light as long as it does not exceed the wattage limits.

Line-voltage fixtures hold R, PAR, metal-halide, and white-sodium sources. White sodium is the most energy efficient. It is 35, 50, and 100 watts with an 85 Color Rendering Index. Use

them!

PAR sources in 60 and 100 watts can be infrared enhanced to produce longer life and brighter light.

Use them for more punch.

COMPARE THE PERFORMANCE
HALOGEN PAR-38 @ 60 WATTS

SOURCE	BEAM	CBCP
infrared	60°	1,300
non-infrared	60°	1,000

Further, a special PAR-20 with a low-voltage source, an internal transformer, and a line-voltage screw base punches even more light.

Metal-halide sources provide 80 or 92 color rendering at 30 or 42K in T, PAR, or ED shapes. They are the low-energy replacement for MR's and can have ultraviolet (UV) filtering.

White sodium produces the least UV. The 35 watt puts out as much light as a 100-watt incandescent with 5 times longer life.

Low-voltge fixtures hold MR, AR, or PAR. Some MR's have infrared technology, some have color constancy, and some filter out UV. MR's are 30K or 47K in very narrow spot to wide flood at 20 to 73 watts. AR shapes are like MR or like a PAR-36 (AR-111).

The aiming angle and the distance away from the display determines the beam width and amount of light. Use point-by-point calculations to predict these unknowns at any aiming angle. Use manufacturers' technical data for 30-, 45-, and 60-degree angles. First, however, identify the manufacturer's definition of zero degrees. Is it perpendicular to the display? Is it perpendicular to the floor? It makes a difference for the 30- and 60-degree data, but not on the 45. Design decisions must be based on accurate information in order to obtain the desired results.

Contrary to what you might think, selection of lighting starts with light-source choice and then moves to fixture choice. Fixture choice only comes first when buying decorative visible fixtures (lighting jewelry) or when retrofitting and the choices of what can be used are limited.

Tracks are useful for display lighting. But, for commercial spaces, they need to be heavy-duty. Multiple-circuit tracks increase lighting possibilities. Recessed tracks make a slick-looking ceiling. Some tracks are already in a trough for distinctive no-source recognition.

Valance light for vertical displays.

Valance and Cornice

Valance and cornice lighting use fluorescent sources. They are excellent for wall displays, clothing racks, and shelves. Install the fluorescent tubes a distance out from the vertical display equal to one-fourth of the distance over which the light is expected to fall. For example, if the display is 4 ft (1.2 m) high, the fluorescent strip should be 1 ft (.3 m) out from the display. (See cornice and valance details in Chapter 5.) Reflectors or a block at a 20-degree angle behind the fluorescent strip can spread the light even farther. Use the triphosphor 1 in. (25 cm) diameter T-8 or compact fluorescent. Both fit a small valance or cove space. The compacts are 4 to 22 in. (10 to 56 cm) long and 7, 9, 13, 28 and 39 watts. Both have a good color (2700K) and render colors well (81 to 87 Color Rendering Index) so essential for successful merchandising.

On the other hand, a valance or cornice board can hide an electric track and incandescent accent fixtures, adding brightness to special features within a wall display. Either fluorescent or incandescent light from vertical displays reflects into the space, giving more general illumination and a spacious feeling.

Luminous Panels

Luminous panels can be vertical or horizontal. They can produce soft backlighting or underlighting for products. Vertical luminous panels backlight shelves. Horizontal panels underlight merchandise, creating a tabletop or shelf surface. Both use fluorescent sources. Both are good settings.

To build a luminous panel, construct a covered box. If the covering is translucent, separate the tubes by a distance equal to one and a half times the depth of the panel. For example, if the panel is 4 in. (10 cm) deep, put the tubes 6 in. (15 cm) apart. Put the tubes well back from the edges at the top and bottom or at both sides, especially if the covering is transparent. Paint the inside of the box off-white to reflect light. (See luminous panel directions in Chapters 8 and 9.)

Additional light can be on the products from the front to highlight and reveal details, if backlighting is not enough. (See Chapter 11.)

A luminous area can be created by diffusing light through translucent fabric. The luminous area needs to be brighter than the surroundings. Fabric spreads the light like a diffuser.

Diffuse light through fabric. Adapted with permission from p. 75 of *Contract* Magazine, December 1979.

Because shelves are always stacked vertically, they are always in view. Thus, shelf lighting should catch the customer's eye, but the source should be hidden. Consequently, choose the source by the amount of light needed to attract attention and/or the amount needed to make the products visible for inspection.

Showcases are enclosed and products displayed have small details. The details need bright light to be seen, but the products can fade and/or deteriorate under such light. A double-edged sword! Thus, choose the source that can fulfill the visual needs, but does it with the least consumption of energy and least product destruction.

For showcases, the least energy-consuming, least-destructive, and brightest source would be a xenon metal-halide with a remote fiber-optic system. Either end-lit with miniature fixtures for accenting or side-lit for a broader distribution of light. The source is in a remotely mounted illuminator box. A single source can light several cases. Such light will not be hot, nor will it contain ultraviolet to harm products. It is the brightest and makes jewelry spectacular. (See Remote Source Lighting in Chapter 20.)

COMPARE THE PERFORMANCE

SOURCE	WATTAGE CHOICES
low-volt tube	$\frac{1}{4}$ to 5
low-volt strip	5 to 18
T-2 fluorescent	11 to 13
T-5 fluorescent	14 to 35
T-5 (HO) fluorescent	24 to 52

For shelf edges or showcases, a low-energy consuming source would be T-2 subminiature, linear fluorescent. They fit glass or wood shelves. They are $\frac{1}{2}$ in. (13 mm) wide at 11 or 13 watts. The fixture is $\frac{3}{4}$ in. (19 mm) deep in 20, 24, 37, and 45 in. (.5, .6, .9, 1.1 m) long. They produce 680 or 860 lumens with a 120 degree distribution and last 10,000 hours.

If more brilliance is required, T-5 fluorescent has longer lengths (2, 3, 4, and 5 ft or .6, .9, 1.2, and 1.5 m). They put out 1,360 to 3,650 lumens, depending upon their wattage. Fixtures are $\frac{3}{8}$ in. (10 mm) deep. Further put, a 3-in. (8 cm) wide recessed fixture above the top shelf with a lens or louver aperture. Thus, the top shelf would be lighted from a concealed source.

In addition, T-5 comes in a high-output version, with 2,000 to 5,000 lumens for the same lengths.

COMPARE THE PERFORMANCE
2 FT LINEAR FLUORESCENT SOURCES

SOURCE	WATTS	LUMENS	LIFE	CRI
T-2	13	860	10,000	80
T-5	14	1,350	20,000	85
T-5 (HO)	24	2,000	20,000	85

Low-voltage tube lights have subminiature incandescent sources. They yield soft light at less than $2\frac{1}{2}$ watts. Tube sources are not replaceable in sealed tubes. They have to be taken down and sent back to the factory. Tubes that open can permit replacing sources. Even with a 20,000-hour life, they do burn out.

Low-voltage strip lights are higher wattage—up to 18—using either 12 or 24 volts with a transformer. The transformer can be remotely mounted, but expect voltage drop determined by how far away. Match the transformer capacity to the wattage needed for the sources minus 20 percent for a safety margin.

**Rule of Thumb for Capacity of
Low-Voltage Transformers**
Transformer capacity = watts of sources + 20%

Low-voltage, strip-light source choices are incandescent halogen or xenon, in clear or frosted. Frosted smooths out the light. Xenon's color temperature is 28K—about 4K higher than non-xenon. Thus, it appears "whiter." It produces less heat and has reduced UV destruction. Some xenon sources operate on line-voltage (no transformer). They are typically more watts—18—with half the heat of halogen and 5 times the life. In a 2-ft (61 cm) high showcase, xenon provides anywhere from 35 to 66 footcandles, depending upon their spacing, 5 to 12 in. (13 to 30 cm) apart. There are many choices for lighting shelves and showcases.

Amount of Light for Stores

The Illuminating Engineering Society of North America recommends that levels of light required for stores be determined by the amount of activity generated. They classify activity into categories—high, medium, and low. High ac-

NUMBER OF FOOTCANDLES FOR STORES

	HIGH ACTIVITY		MEDIUM ACTIVITY		LOW ACTIVITY	
Showcase and Wall Displays	100	(1,000 lux)	75	(750 lux)	30	(300 lux)
Feature Displays	500	(5,000 lux)	300	(3,000 lux)	150	(1,500 lux)
Circulation	30	(300 lux)	20	(200 lux)	10	(100 lux)

Table adapted from IES Lighting Handbook, 1987 Application Volume.

tivity is easily recognizable merchandise, rapid viewing time, and impulse buying. Medium activity is familiar merchandise, longer viewing time, and thoughtful decisions. Low activity is unfamiliar merchandise, with assistance and time required for viewing and decisions.

Energy-efficient light sources, incandescent, fluorescent, metal-halide, and high-pressure sodium, have been developed over the past few years. Since 60 percent of the energy consumed in retail business is for lighting, it is more important than ever to utilize these more efficient sources. They do not have to decrease the quality of lighting. In fact, they actually can increase it with better color-rendering properties, longer rated life, and higher lumens per watt. Use them.

Color of Light for Stores

Color sells. The visual appeal of color is undeniable. Color of products can be intensified or diminished by light. Four factors play a role—amount, spectral distribution, color temperature, and color-rendering ability of light.

If the amount of light is too low for a particular source to render colors well, colors become dim (like an underexposed photograph). If the amount is too high, the colors become washed out (like an overexposed photograph).

Further, the color of a product cannot be seen unless the color of the product is present in the light source—its spectral distribution. But, spectral distribution curves do not help the decision maker to know what will be seen. Hence, the illumination scientists developed scales to help predictions about color. The Kelvin temperature scale (K) indicates the coolness or the warmness of the light from a source and the Color Rendering Index (CRI) indicates how the source renders surface colors. (An additional scale—Color Preference Index, sometimes called the flattery index—indicates how pleasantly a source renders colors.) None of the scales are perfect. Use them with caution and understand their shortcomings for your purposes.

KELVIN TEMPERATURE

Since color of light is hard to describe, scientists use Kelvin temperatures (K) for indicating the coolness or warmness of light. However, the Kelvin temperature scale is only accurate when referring to something heated from cold black through red hot, white hot, and on to brilliant blue hot. Only candles, sunshine, or incandescent light can be accurately measured with Kelvin degrees. Nonetheless, for convenience, Kelvin degrees are also used to describe the apparent color of light from discharge sources—fluorescent, metal-halide, sodium, mercury, and cold cathode. The degrees give an idea of how warm or how cool the light appears. The scale runs from 1,900 Kelvin (K) to 26,000, often shortened to 19K and 260K. The lower the number, the warmer the color. The best way to remember the Kelvin scale is to think of an outdoor scene at sunset—with the sun setting, the sky low on the horizon is red (low in Kelvin degrees); the sky high above is blue (high in Kelvin degrees). The memory device is: low temperature—reds low in the sky; high temperature—blues high in the sky.

Use Kelvin temperatures to realize how cool or how warm the light will appear in the space. If the space is to be dimly lit, use warm sources. If the space is to be brightly lit, use cool sources, particularly if the interior colors are to be warm. Do not use super-cool sources with totally cool interior colors, nor super-warm sources with totally warm interior colors. Balance interior color schemes with the opposite temperature in sources.

Never mix different Kelvin temperature sources in a visible fixture, such as cool-white and warm-white sources in the same direct-light ceiling fixture. The effect is horrid when looking directly at it; the effect is unnoticed when looking at the merchandise. Colors of light blend on surfaces, but not at the visible source.

COLOR RENDERING INDEX

The old standby, incandescent, is around 2700 to 2900K. Consequently, if a new source, like super high-pressure sodium is announced at 2700K, the temperature is a lot better than high-pressure sodium at 2100K. However, do not be fooled when Kelvin degrees are similar. If the sources are different types, their ability to render colors may be very different. The Color Rendering Index (CRI) was developed to predict how well surface colors will look under a particular source. The highest rating is 100. The lowest rating is 20. Warm-white deluxe fluorescent has a CRI rating of 77; daylight fluorescent has a rating of 75. But, they render colors differently because their Kelvin degrees are different (3000K and 6250K respectively). Compare indexes only if the sources have Kelvin degrees within 300 degrees of each other. Both the Kelvin temperature and the Color Rendering Index must be used in conjunction with each other and used cautiously. The comparisons are not absolutes and are based upon our subjective impression of what appears normal.

DICHROIC LIGHT

Dichroic PAR sources selectively transmit one color reflecting back all others. The one-color light enhances surfaces of the same color. Surfaces of other colors are not affected. For example, blue dichroic sources are favorites for jewelry stores to heighten blue backgrounds for glittering jewelry. If the blue dichroic light happens to fall on another surface, no light is seen. The enhancement is subtle but profound. Since the source is usually hidden, the viewer wonders where the richness of color is coming from. (See Tinted Light in this chapter.)

Displays in Windows

Displays in windows require light. The amount required depends on many things. Are the displays viewed during the day or at night? Are they viewed in competition with other windows? Do the windows receive reflections from other light sources, such as streetlights?

Rule of Thumb for Lighting Display Windows
Feature displays require high illumination during the day when the outside light is bright and at night in a highly competitive area when external lights can reflect.

A photosensor and a dimmer can raise the footcandle level in the window when daylight gets brighter, and they can lower the level at night when the higher level is not needed. Consequently, these devices can make sure that the window is bright enough to have the potential of attracting attention without wasting electricity.

The Illuminating Engineering Society divides window lighting into general or feature displays, and they recommend these amounts:

NUMBER OF FOOTCANDLES FOR STORE WINDOWS

	DAYTIME	
General Display	200 (2,000 lux)	
Feature Display	1,000 (10,000 lux)	
	NIGHTTIME	
	Highly Competitive	Noncompetitive
General Display	200 (2,000 lux)	100 (1,000 lux)
Feature Display	1,500 (15,000 lux)	500 (5,000 lux)

Some display windows have no back wall, and anyone can see through them into the store. They are called see-through displays. External and internal brightness affects the amount of light required in the see-through display. External light (daylight, mall lighting, or streetlighting) enters these windows, adding to or detracting from the window display. Likewise, internal light affects the attention-getting quality of the see-through display. Consequently, the brighter the interior, the brighter the display needs to be. See-through displays also permit outside light to enter the store, adding to or distracting from the interior brightness. They can reduce the store's dependence on electric light, but they require careful planning to be balanced, glare-free, and successful.

Light Sources

Many light sources are available for window or store displays. Light source catalogs are the best references. Catalogs use a system of shorthand to identify the source. For example, Q75MR16/NFL means a quartz, 75 watts, MR-16 incandescent-type source in a narrow flood distribution.

Choose the source that gives the beam spread that just covers the area to be accented. Accordingly, the accented area will have the greatest amount of light and use the lowest amount of watts. In many cases, the choice will

---

Table adapted from IES Lighting Handbook, 1987 Application Volume.

Features in a display window
require higher illumination.

A see-through display window
permits outside light to enter.

be low watts in either low voltage (MR-16 or PAR-36) or in line voltage (T-10 white sodium or PAR-20). Remember low voltage does not always mean low wattage; it means more brightness and longer life at the same wattage.

COMPARE THE PERFORMANCE
OF LOW- AND LINE-VOLTAGE PAR

SOURCE	WATTS	BEAM	CBCP	LIFE
12-volt:				
PAR-36	50	8°	11,000	4,000
120-volt:				
PAR-16	50	10°	8,250	2,000

Different light sources have comparable but not exact same beam spreads. They do not have the same life spans. Long-life sources are preferable for window displays!

MR sources are halogen, low-voltage. They accent displays well. MR-11 and MR-16 wattages range from 20 to 73 and beam widths from very narrow spot to wide flood. MR-16's with infrared technology are the brightest. Some MR-16's are color stabilized with lower UV output. One MR is 47K, illuminating colors like daylight. (Others are 30 and 32K.) MR life is 3,000 to 5,000 hours.

COMPARE THE PERFORMANCE
LOW-VOLT, SCREW-TERMINAL PAR-36

WATTS	BEAM	CBCP	LIFE
25	5°	19,700	1,000
35	5°	25,000	4,000
50	5°	40,000	4,000

PAR sources are either line- or low-voltage with medium screw, side-prong, or screw-terminal bases. Line voltage with infrared technology is brightest; low-voltage PAR-36 is also bright.

Halogen sources, both line- and low-voltage, have the added punch of whiter light and enhance colors vibrantly. In addition, halogen sources frequently provide the same amount of footcandles at a lower wattage.

Choose source with the beam spread that just covers
the area to be accented.

Adapted with permission of Litecontrol.

life, and more central beam candlepower (CBCP) for a little more electricity, the 35-watt source is it.

Within the same wattage (60 or 100), line-voltage halogen PAR's offer spots to wide floods. Thus, beam widths could be changed as often as the displays change. Lots of choices!

A special line-voltage PAR-20 and -30 with a low-voltage beam at 20 and 35 watts can produce bright light. It has an internal transformer.

Also, a remote-source, end-lit fiber-optic system could provide bright light with no heat and less UV—excellent for irreplaceable displays.

COMPARE THE PERFORMANCE
OF LINE VOLTAGE
FOR WASHING WINDOW DISPLAYS

SOURCE	WATTS	KELVIN	LUMENS	LIFE
fluorescent:				
long compact	55	30-41K	4,000	14,000
compact	42	30-41K	3,200	10,000
white sodium:				
T-10	50	27K	2,500	10,000
ED-17	50	27K	2,000	10,000

For washing large areas with light, clearly long-

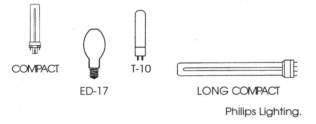

COMPACT   ED-17   T-10   LONG COMPACT

Philips Lighting.

compact fluorescent (TT-5) lasts the longest and produce the most lumens. The long-compact fixtures are rectangular; compacts are round apertures; sodium fixtures are either.

White sodium produces light similar to incandescent in color and brightness using 33 percent less energy and lasting 10,000 hours. It is 27K at 85 CRI in 35, 50, and 100 watts. It enhances texture and creates highlight; fluorescent does not.

COMPARE THE PERFORMANCE
OF LOW-VOLTAGE NARROW FLOODS
FOR WASHING WINDOW DISPLAYS

SOURCE	WATTS	BEAM	CBCP	LIFE
stable-color, infrared, MR-16	50	32°	6,000	4,000
halogen, infrared, PAR-30	60	29°	1,100	3,000
halogen, PAR-30	60	29°	800	2,500

Among incandescent sources, those with infrared enhancement are the best.

Clients who object to higher initial costs of fluorescent, low-voltage, or sodium might be persuaded that after initial payback (1 to 2 years), the energy saved is a cash flow for the life of the building or the life of the business, whichever comes first.

power for harm

Light can harm products. Light sources radiate heat. Heat can be uncomfortable for customers and employees. Often, the heat needs to be reduced with air-conditioning. Moreover, it has an effect on products causing perishing and discoloration. Some products melt (candy). Some products discolor (meat, both fresh and frozen). Perishing can occur at any time. Sometimes discoloration occurs in hours; sometimes in days. Discoloration of food is caused by changes in the product, including bacterial growth in the warmth of the light. Both melting and discoloration reduce sales potential.

Lower wattage means less heat. Fluorescent and low-voltage incandescent sources are cooler than standard and halogen incandescent, because of lower wattages. However, some PAR's are designed to produce a cool beam. These PAR's filter out two-thirds of the heat, sending it through the sides rather than forward with the light. Specify these PAR's when a point source is required and reducing heat or increasing the amount of light with no additional heat is desired. For example, when competition is keen, increase the light in the window to attract attention, but keep fading the same with higher-wattage cooler source. For example, when displaying food, keep the amount of light the same, but reduce radiant heat with a cooler source.

In addition, light sources radiate ultraviolet, the invisible light that fades products. Most materials fade. Fading usually occurs between 50,000 and 70,000 footcandle hours (footcandles multiplied by time). For example, if a product with unstable dyes is displayed at 500 footcandles for 100 hours, it can fade. Pale colors often fade faster than more brilliant colors. Fading is more apparent when some parts fade and some do not because they are away from the light. For example, neckties folded in display cases can fade on the edges and on the top. Products should be rotated out of the light every 10 days to control fading. Fading is less apparent if the whole product is uniformly exposed to the light. Products should be spread, not stacked, in and on cases. All light sources cause fading. The hotter the source, the more the fading. Keep light bulbs at least 9 in. (23 cm) away from products. (See Chapter 11 for more details on fading.)

power for matching colors

Many businesses require seeing or matching colors correctly: medical-testing labs, printers, hairdressing salons, fabric shops, museums, clothing stores, art studios, and others. Sometimes matching colors is critical (medical); sometimes less critical, but important nonetheless (printing). Color-matching fluorescent sources can match for different needs. For critical color matching, fluorescent sources at 7,500 Kelvin degrees (75K) are required. Their light is like north-sky light. They render colors to meet the standards of the American Society for Testing Materials. For less critical color matching, sources at 50K with a high color rendering index (CRI) are acceptable. Their light is like natural daylight. They meet standards for graphic art color matching and for viewing photographic transparencies. These sources have a broad spectral distribution composed of a curve that matches natural light at that Kelvin temperature.

For some businesses where color matching is desirable, but not critical—jewelry, fur, cosmetics, hairdressing—skin tone appearances are part of the equation for sales. Thus, 30 or 35K sources with high color-rendering indexes would flatter faces better. T-8 sources offer a broad range of CRI from 95 down to 75, both at 30K. Their spectral distribution is a series of spikes, not like natural light at that Kelvin temperature. The eyes do not see these spikes, but critical color matching would be disadvantaged by such light. When choosing, be mindful of the effect that the color of light will have on the colors of the space, particularly pale hues. (See Apparent Tint Given to Surface Colors in Chapter 2.)

Color matching is important in other circumstances. The business of architecture and interior design is partly based upon color matching. Architects and interior designers want colors chosen in sample size to appear the same in the installation and want to have custom-dyed goods appear correct to the client. Often, it does not happen. There are many reasons. Surface characteristics, color adaptation, influence of adjacent colors, and lighting are among the reasons. Surface characteristics (dye and fabric composition) should be the same in the sample as in the actual goods. Color adaptation can not be controlled by architects and designers—nor anyone else. At the same time, design professionals rely on experience and memory about the influence of adjacent colors. Memory might not always be correct. But, design professionals can match samples under the same light source that will be used in the installation. This practice will prevent their worst nightmare of metameric match—when two samples match under one light source but do not match under another. The color shift is astounding; it can be avoided. At the least, build and use a light box.

See for Yourself: Will the Colors Match?
1. Build a light box of two 20-watt T12D fluorescent sources and two 40-watt silvered bowl incandescent sources, separately switched.
2. Colors that match under these two light sources should match under any source.

Reprinted by permission of General Electric.

Carpet and fabric manufacturers have had to deal with these light-induced color discrepancies, particularly when matching (or attempting to match) custom-dyed products. It has been such a problem that the manufacturers are printing literature about lighting and suggesting lighting changes to solve color mismatch, making colors appear correct.

As lighting technology advances and more and more sources are produced with different color temperatures, color-rendering abilities, and color-preference effects, color-matching problems will increase and pretesting for matching will be necessary.

power for productivity

In both shops and offices, lighting has the power to create a productive work environment. When the light level is low, visual efficiency is poor and work can slow down. The cost of providing the recommended light levels is typically only one percent of employees' salaries. Good light is good for business. Businesses that provide a service evaluate their light by its effect on people and their tasks. Illuminate the task first and the space second. Sometimes a task fills a space. In such cases, room lighting needs to be uniform to obtain even task lighting. The amount of light required for a task depends upon how difficult the task is, how important it is, and the ability of the employee to see. The more difficult the visual task, the more important the task, .and the older the employee, the more light is required. The requirement for reading ranges from easy at 30 footcandles to difficult at 150 footcandles. (Consult the **IESNA** Lighting Handbook, Reference and Application, 1993.) Tasks, also, need glare-free and shadow-free light. Consequently, task light must not come from above or in front of the task. (See Chapter 19.)

Uniform and Nonuniform Lighting Design

Design lighting for small commercial spaces by deciding whether the whole space must be uniformly lighted or not. Uniform lighting spreads the same amount of light from wall to wall. Why light the whole office at 50 footcandles when much of the light falls on the floor? Uniform lighting is not always the best. Nonuniform lighting makes some areas brighter than others, usually determined by task requirements or desired effects. Nonuniform lighting can provide 50 footcandles at the desk and spread 10 footcandles elsewhere. This practice can be more energy-efficient, more interesting, and more adaptable.

SOURCES

Sources for uniform lighting are either direct or indirect. Direct light is usually produced by fluorescent sources in average-height spaces, and metal-halide, sodium, and deluxe mercury sources in high-ceilinged spaces. Careful planning is required to get smooth light throughout the space with no hot spots (areas of bright light) anywhere, including the ceiling. Smoothness is determined by fixture design, placement, and bulb wattage. If the fixtures are well engineered to put the light down without brightness at the source, correctly positioned, and correctly lamped (have the correct bulb type and wattage), the light will be distributed uniformly.

Indirect light is produced by either metal-halide, deluxe mercury, sodium, or fluorescent sources. Metal-halide, mercury, and sodium cost more to purchase and install than fluorescent, but use less electricity. Therefore, the initial cost is offset by lower long-range cost. Indirect light is bounced from somewhere else (the ceiling or a wall) before it falls into the space. Indirect light can supply illumination for either uniform or nonuniform lighting.

Nonuniform lighting sources can be all direct or both direct and indirect. Nonuniform direct light is produced by point sources (regular or low-voltage incandescent and high-intensity discharge). Indirect light can be produced by point sources (high-intensity discharge) or linear sources (fluorescent). When used as direct light, point sources create highlights and shadows; when used as indirect light, they can spread illumination over a broad area. Linear sources do not effectively create highlights and shadows.

Uniform lighting.

Nonuniform lighting.

FIXTURES

All fixtures for direct light are positioned to put light down. Direct lighting fixtures can be on the ceiling, recessed, surface-mounted, or suspended. On the other hand, indirect lighting fixtures are positioned to put light on a large architectural surface—a ceiling or wall. They can be ceiling-hung, wall-mounted, furniture-integrated, or freestanding. Be aware that furniture-integrated and freestanding indirect fixtures are often classified as furniture for tax purposes. Therefore, they can be deducted at the accelerated rate of furniture rather than as building equipment. Likewise, such fixtures do not require disturbing the fire-resistant ceiling of a structure, and a lower fire insurance rate is possible. Many considerations are required to decide which system is the best. They include aesthetics, cost of installation, tax break, insurance rate, cost of maintenance, cost of electricity over time, and available funds. The decision is complex at best, and detailed specifications are required for bid comparisons.

LIGHT LEVELS

In uniform lighting, the whole space is illuminated to the level required for the task. In non-uniform lighting, the task area is illuminated to the level of the task, and the general space (ambient light) is lighted at a lesser or greater amount as required. The whole room is not lighted to the same level, and less energy is consumed. Ambient light can be aesthetically pleasing and define the space. It can provide safe passage and enhance the image of a business. Some areas of ambient light can be brighter than the task area. The surfaces can be up to 10 times brighter. Bright ambient light can come from a large surface washed with light—wall-washing, cove lighting, and uplighting. Some remote areas can have surfaces softly lighted with as little as one-tenth the brightness of the task. Soft ambient lighting can come from accent, downlighting, and art lighting.

OPEN-OFFICE LIGHTING

Open offices (large offices with many desks) present lighting problems, as reported in an analysis of five open-office lighting systems by Noel Florence (see the Bibliography). Open offices have partitions and furniture that can be changed, usually needing but not getting light fixtures repositioned. Lighting for open offices can come from ceiling-hung sources (direct) yielding uniform illumination, from task-located positions yielding nonuniform illumination, from widely spaced ceiling sources and furniture (undercabinet or shelf-mounted) sources (both direct), or from furniture-mounted or freestanding uplight sources (indirect) for the space and furniture-mounted (direct) for the task. Shelves and cabinets overhanging furniture can block the light from the ceiling. Also, direct light suspended from cabinets and shelves for tasks can cause reflected glare and poor visibility. Noel points out that office lighting systems should be evaluated for energy effectiveness, which includes energy input, visual comfort, and relative visibility at the task location. He states that spacing of the task locations also affects the lighting effectiveness. Therefore, a single measure of watts per square foot or square meter (building watts) is not the only appropriate measure. Likewise, watts per task location (people watts) should be considered. These two measures should be used to design the best system for open-office lighting.

Calculations and design must be carefully done for open-office lighting. Spacing criteria do not work if there are vertical barriers (partitions), particularly those with cabinet or shelf overhangs. Reflectances are not constant because some surfaces are not only a color with a specific reflectance, but also in shadow, diminishing the reflectance value. Consequently, interpreted calculations based on coefficient of utilization are often inaccurate. They usually indicate that the task will receive more light than it will. Steven Orfield (see Bibliography) suggests that the most important variables for open-office lighting are:

- distance to the task
- orientation of the source
- intensity of source in task direction.

Further, he suggests that solutions be defined in these terms before products are considered. He calls this method—coefficient of application. Use it rather than coefficient of utilization.

COMPUTER LIGHTING

Computers increase productivity and at the same time increase lighting problems. The monitor screen (visual display terminal, VDT) is glass and reflects sources and bright surfaces. Hence, the glare-producing zone (zone from which a bright source would cause glare) is in a different location with computer tasks (behind the desk) than for pencil/paper tasks (in front of the desk). (See Chapter 19.) Consequently, uniform ambient lighting is not suitable. Nonuniform task lighting is required. But, light must come from the correct angle—not above 60 degrees from nadir. Also, the light needs to be the correct amount to show contrast between the image being read in hardcopy and its background and at the same time not wipe out the monitor's contrast. Likewise, light must be sufficient for the age of the worker, permitting the speed and accuracy required. Older workers require more light to get sufficient contrast, but dull monitors require less light to enhance minimal contrast. What a dichotomy! Further, hardcopy and monitors should be the same brightness. Otherwise, the user will get visual fatigue looking from one to another.

In rooms with multiple computers, the difficulty of getting light at the correct angle is enormous since computers face every which way. In some of these rooms—stock exchanges, traffic control rooms, etc.—the walls include many monitors and angled banks of control equipment. None should receive glare. Also, acceptable uniformity of these vertical surfaces is mandatory, both side to side and top to bottom.

Rule of Thumb for Acceptable Uniformity or Luminance Ratio of Vertical Surfaces
side to side	1:1.5
top to bottom	1:5

Visibility is the key word. How visible will the task be? Answer before installing.

Fixture Determination

In commercial spaces, the type of ceiling determines the options for ceiling-fixture installation. Ceiling construction is usually one of the following types—wood ceiling joists, suspended grids, open bar joist, or concrete and steel. Wood ceiling joists, suspended grids and bar joist types accept recessed or surface-mounted fixtures. If the ceiling is concrete and steel, the fixtures must be surface-mounted, unless they are set in before the concrete is poured.

The decision of what fixtures to buy depends on many decisions and trade-offs—color of light, initial cost, operating cost, ease of maintenance, potential noise (from ballasts),

Getting the correct angle of light
in rooms with multiple computers
is difficult; a reflected-light
ceiling-coffer system helps.

payback time, and aesthetics. Choose wisely; well-engineered fixtures pay you back over time. After determining which fixtures are possible, develop the reflected ceiling plan. Be sure to pay attention to the pattern the fixtures make on the ceiling. They are unmistakably in view, and lighting design for offices is composed of more than fixture determinations.

BALLAST DETERMINATION

Fluorescent and HID fixtures must have ballasts. A ballast is used to start and operate the source. After starting, the ballast limits the electricity to the range acceptable, as determined by the manufacturer. Without this electrical restraint, the light source would burn out.

Fluorescent sources are classified by their starting characteristics, either needing or not needing a starter. Instant start and rapid start do not need starters. Rapid start are used the most. They have a transformer in the ballast to get them going. They are the only ones that can be dimmed. Instant start can be distinguished by single pin ends. Instant start use high voltage to get going. Preheat sources need a starter. A switch initiates heat for starting. Ballasts must be compatible with the source. Thus, ballasts are often classified by these starting characteristics also.

Two different types of fluorescent ballasts are made. They are magnetic (core-coil and reactance) and electronic (solid-state). Core-coil ballasts are used for rapid-start and for instant-start fluorescent sources. Some save energy, operate cooler and last longer. Reactance ballasts are used for low-wattage, preheated fluorescent sources and require an external starter.

The second type of ballasts, electronic, is also called solid-state. They are used for both fluorescent and HID sources. They operate the source at a high frequency, increasing source efficacy and eliminating the fluorescent flicker that bothers some people. However, the frequency is in the radio-frequency range and can affect any electronic equipment. Consequently, the Federal Communication Commission requires a radio-frequency interference (RFI) device. Electronic ballasts are more delicate than magnetic. They can be damaged by electrical surges, but some manufacturers build in transient surge protection.

Energy consumption for ballasts differ. A two-source 40-watt fluorescent fixture and core-coil ballast consumes 16 watts for a total of 96 watts, including the sources. On the other hand,

electronic ballasts cut the wattage. They do the job for a total of 70 watts, including 15 for the ballast and 55 for the two sources. (Light source watts are reduced by a cooler ballast and cooler operating sources, translating into less total watts consumed.) The reactance ballast consumes 18 watts and demands 80 for the two sources with a total of 98.

ELECTRICITY USED BY FLUORESCENT BALLASTS
USING TWO 40-WATT SOURCES

BALLAST TYPE	SOURCE WATTS	BALLAST WATTS	TOTAL WATTS USED
Core-coil	80	16	96
Electronic	55	15	70
Reactance	80	18	98

Reprinted with permission of General Electric.

Ballasts have little tolerance for variability. They are adversely affected by temperature and by wrong light sources. Many ballasts are thermally protected against internally overheating (indicated by a class P). A thermal protector that turns off the ballast if it overheats is required by the National Electric Code. Ballasts are designed to start sources at certain minimum temperatures. Most are designed for 50 degrees Fahrenheit starting. Others are designed for zero or minus 20. Ballasts are sound rated, A to D, indicating quietness. A is the quietest. Ballast hum is a symphony most people do not want to hear.

BALLAST LABELS

Standards for ballasts are set by American National Standards Institute (ANSI) and a CBM label (Certified Ballast Manufacturer) indicates that they meet ANSI specifications. A UL label (Underwriters Laboratory) indicates that United States safety criteria standards are met (CSA for Canada).

Labels indicate information about ballast standards. Labels show voltage and line current requirements. Labels show approval for meeting thermal protection requirements—class P. Labels show "no PCB," if no carcinogenic polychlorinated biphenal is used. Labels show type of light sources UL listed for the ballast. Labels show "HPF" (high-power factor), indicating that the ballast delivers more energy and consumes less. (Some utility companies charge a penalty for installations using low-power factor ballasts, thereby causing a greater and unnecessary electricity demand. Some areas have codes requiring the HPF ballasts.) Finally, labels show a sound rating, "A" through "D." Ballasts have standards of performance; choose the best.

Maintenance

Providing sufficient light is only the beginning. Maintenance is required and includes bulb replacements, fixture adjustments, and cleaning. Many an appropriate and even spectacular lighting design has been ruined by improper bulb replacements or lack of fixture adjustment. (How often have you sat in an expensive hotel meeting room and could not see the speaker because the ceiling fixtures—the ones with bulbs—were facing every which way but the direction intended—on the speaker?)

Poor maintenance does not happen just in hotels; it happens in offices, stores, and even homes. In hotels and offices, everyone waits for the maintenance staff to fix the fixtures. In stores, poorly lighted displays should cause the owner to pay attention. In homes, fixture maintenance is usually thought of as bulb replacement only. But, any fixture requires cleaning. Many installations providing ample light have been reduced to insufficient light if the fixtures have not been cleaned. Dust must be removed. It can cause a 20 percent loss of light, impeding the necessary work, particularly if the initial level was minimal to conserve energy.

To further complicate matters, a light bulb produces less light as it ages—12 to 15 percent less. Consequently, the amount of light is decreasing all the time until the fixture is relamped (new bulbs put in) and recleaned.

Maintenance means organizing the maintenance staff, training and supervising them to relamp, adjust, and clean the fixtures. It also means cleaning, repainting, or rewallpapering the walls and ceiling to reflect the greatest amount of light.

Conference Rooms

Conference rooms are spaces where information gets disseminated and where productivity is important. They are multiused spaces with multiple lighting demands. Their uses vary:

- Informal meetings with difficult visual tasks (studying corporate records, some as hardcopy and some on a computer monitor).
- Formal meetings with easy visual tasks (review-

ing an annual report with overhead projection of graphics).
- Entertaining corporate guests with no visual tasks as such.
- In-house corporate training with complex visual tasks (video projection and speaker needing light on reading materials and on his/her face for audience visibility).

The surfaces requiring light are: tabletop, speaker's face, speaker's notes, faces around the conference table, walking space, hardcopy displays, audio/visual control equipment if within the space, and backwall if video recording (broadcasting or taping). Audio/visual communication methods differ and have different lighting requirements:

- Projection methods—video, slides, overhead, and motion pictures.
- Display methods—hardcopy, electronic blackboards and monitors (computer and video).
- Recording methods—videotaping or broadcasting (teleconferencing).

First, with projected audio/visual methods and note taking, light on the tabletop could be 3 footcandles. Whatever amount, it should be confined to the tabletop. If fluorescent lights are used, choose reflectors to aim the light where needed—either directly down when avoiding spill light, or asymmetrically when highlighting. Also, choose a fluorescent with a color temperature compatible with other sources in the space. Color temperatures should not vary more than 150K if falling on different surfaces when video recording. (See Sonnenfield in Bibliography.)

Second, with the use of electronic communications (electronic blackboard, teleconferences, electronic mail, and facsimile transmission), meetings originating in one location can have input from other locations instantly and transfer information. The lighting should not wash out communication methods. Consequently, screens (front- or rear-projection) should not have additional light directly falling on them. Rear-projection screens are not as sensitive to additional light as front-projection screens. The screen brightness with projection should be 5 times as intense as the screen without projection. For clarity and fine details, the screen could be 100 times as bright with projection.

Surfaces needing light are:
speaker's face and notes,
faces at tables, tabletops,
and graphic displays.

Luminance Ratio of Screen with Projection Illuminance to Screen with Room-Only Illuminance

5:1

Similarly, monitors (computer or television) should not have additional light (direct or reflected) on them. The monitors need to be adjusted for the space to have at the least a 10:1 contrast ratio of the monitor image to its background.

Minimal Contrast Ratio for Monitor Image to Its Background

10:1

Third, when video recording, the better the quality the camera, the less light required to illuminate the scene. The lighting must keep the person from fading into the background. The background cannot be white. The camera can not register well against a white background. Use theater lighting techniques: a strong front-light (coming from left or right at 45 degrees out and 45 degrees up), a softer fill light (coming from the opposite side) and backlight (highlighting the outlines of the head and shoulders). This is the well-used Stanley R. McCandless' method of theater lighting. (The difference between theater and architectural lighting is that in theaters the people are lighted; in architecture the structure and its contents are lighted and reflect on people.)

Equally important, speakers do not speak from the same place. Some speak from the podium and some speak from the seat at the head of the table. Some speak while walking around the room. As a result, lighting the speaker is like lighting a ballet—a leaping and turning target. Light for the speaker's face must make the face appear brighter than the background. (Dark skins require a lot of brightness.) It is impossible to visually focus on a drab target for very long. Other brightness attracts our attention. If the ceiling happens to be brighter, people will look at the ceiling, rather than the speaker.

Therefore, luminances must be controlled. Sometimes created; sometimes diminished.

Luminance Ratio for Speaker's Face to the Background
2:1½

Light for the speaker's face when video recording needs to be front-, fill, and backlight. The frontlight can be 3 times brighter than the fill light. The backlight can be about the same brightness as the front- and fill lights together. Like the theater, the light sources for the front- and fill lights should be direct point sources to create shadows and show details. Take advantage of the best technology—lenses, reflectors, and sophisticated beam spreads.

Ratios for Theater Lights
Frontlight to Fill Light
3:1
Backlight to Front- and Fill Light
1:1

Ratios reprinted by permission of Illuminating Engineering Society. (See Sonnefield in Bibliography).

In addition, speaker's notes need to be illuminated as reading matter—not below 20 footcandles. The light must be confined so that it will not cause unflattering shadows on the face. Likewise, hardcopy displays—flip charts, white and green chalkboards, magnetic displays, and posters—need to be illuminated as reading matter. Faces around the table should not be dimmer than their background or lighted in silhouette. In silhouette, the person would appear anonymous and expressionless. The walking space—the carpet leading to the door—needs to be lighted at no less than 2 footcandles. Any audio/visual control equipment contained within the conference room must be lighted with confined light so that adjustments can be made or failures fixed without altering the room lighting.

Dimmers—manual and programmable—can organize the multiple lighting demands of conference rooms. They can preset lighting amounts for the various meeting types and visual tasks. However, dimmers must not put out an audible hum or a radio frequency to interfere with the electronic equipment. The better ones do not. All in all, lighting a conference room is complicated. Nonetheless, design the lighting to accommodate the state-of-the-art audio/visual equipment. State-of-the-art equipment will be standard equipment tomorrow. Redesigning the lighting later is costly. Prepare for the best in telecommunications now. (See Bibliography under Anderson, "Boardroom Shakeup.")

power of glare

Reflected Glare

Glare can be reflected from shiny surfaces. When it reflects and obscures visual information, it is called veiling reflections. In offices, veiling reflections occur on glossy printed pages and on TV monitors. In commercial spaces, veiling reflections occur on glass showcases, transparent packages, windows, glass-covered art and glass-covered directories at elevators. Veiling reflections blur visual information and cause inconvenience with loss of productivity.

In addition, glare can be reflected from finish materials: mirrors, glossy paint, polished marble, and gleaming metal on walls or floor. Such reflections give inaccurate visual cues. For example, in a Seattle bank, the polished marble floor reflected a grid ceiling with light sources behind it. Unfortunately, the reflection of the lighted grid made the polished floor look like steps going up or down. It could confuse those who entered.

Likewise, shiny surfaces can reflect a mirror image of the space which looks like additional space. Such mirrored reflections can confuse people. For example, an interior design instructor related how he designed a restaurant and covered large square columns with mirrors, hoping to hide them. He did. But, he also hid cues to the real space. People bumped into them. He was called back to redesign. Be careful of both reflected glare and mirror reflections.

Direct Glare

Glare can come directly from fixtures. This type of glare is concentrated bright light. It draws people's attention. It is hard on the eyes and can

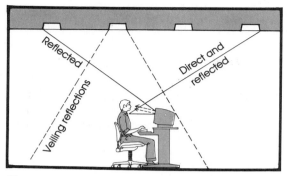

Glare

be fatiguing. Ceiling fixtures and other visible fixtures with bare bulbs or translucent diffusers are the primary offenders, particularly if seen against a dark background. Therefore, ceiling fixtures should not be glaring, particularly fixtures over 4 sq ft (.037 sq m), including luminous panels and artificial skylights.

Manufacturers of fluorescent fixtures publish visual comfort ratings. It is called Visual Comfort Probability (VCP). VCP is an evaluation of direct glare in a room lighted by specific luminaires and judged by people seated in the worst possible place (centered in the back of the room). The rating is the percentage of the people who describe the lighting as comfortable.

Aperture coverings for fluorescent fixtures have varying degrees of visual comfort. Parabolic louvers are considered the most comfortable (91 VCP). Other louvers and lenses are considered less comfortable (50 to 90). Diffusers are considered the least comfortable (40 to 50).

Rule of Thumb for Visual Comfort Probability
VCP ratings should be 80 or more for large spaces. VCP does not apply to spaces less than 20 ft (6 m) in both directions. The luminaires are essentially not in view.

On the other hand, visual comfort from a point source (incandescent and high-intensity discharge) is not usually rated. Further, the visual comfort is controlled not only by the type of aperture and covering, if there is one, but also by the engineering and materials of the fixture. For example, some fixtures with cones around the aperture create glare by an unwanted reflection of the light source. Baffles around the aperture do not reflect the source. In general, well-engineered incandescent fixtures control glare

better and produce more light (sometimes as much as 400 lumens more per fixture). The VCP will be the greatest when the aperture is not overly bright. Hence, the least comfortable is a fixture with the face of the light source positioned at the aperture.

Louvers, Lenses, and Diffusers

Fixtures often have a covering over the aperture—louver, lens, or diffuser. The covering affects the Visual Comfort Probability (VCP), the intensity, distribution, and aesthetics. Choosing aperture coverings can be complex. Besides comfort rating, the qualities of color stability, cleanability, fire rating, cost, and light transmission need to be considered.

LOUVERS

Louvers conceal the source and influence light delivery by their shape, position, and finish. Louver shapes are cells—cubes, cylinders, or baffles. The sides of the cells are either straight or a portion of a parabola. The straight-sided louver is called eggcrate. It is the least expensive and the least engineered.

The parabolic-sided louver for fluorescent fixtures is either a single or double wedge, with the intent to direct the light down with low brightness at the aperture and high brightness on the work surface. Louvers establish a 45-degree cut-off angle (and an equal shielding angle). Light that strikes the louver at more than 45 degrees is redirected ultimately into the 0- to 45-degree zone by the parabolic louver. In retrofits (putting new parts into an existing luminaire) with a ceiling full of fluorescent fixtures, when low-VCP lenses are replaced with high-VCP parabolic louvers, the room aesthetics often change. Less brightness is evident on the ceiling, and colors in the space seem to darken. Be aware that such changes affect overall distribution and amount of light and, therefore, can affect perception.

Louvers are made of polystyrene, acrylic, or aluminum. The colors are: silver, gold, copper, bronze, white, or black with different reflective percentages. They have different finishes—specular, semispecular, and satin. Specular-parabolic louvers reflect light down. Semispecular finishes reflect wider with a 5 percent lower VCP rating and higher Spacing Criteria (SC). (Spacing Criteria establishes how far apart fixtures can be placed in order to have uniform lighting. Multiply SC times the mounting height to get spacing for luminaires. Do not be fooled; the criteria guarantees relatively uniform light, but does not guarantee at what intensity.) Finally, satin finishes scatter light broadly with a little lower VCP and an even wider criteria (up to 6).

LENSES

Lenses have specific functions. Some lenses alter the spectrum of the light—an ultraviolet filtering lens. Ultraviolet-filtering lenses remove harmful ultraviolet that can destroy fine art.

Concave lens.

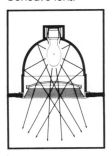

Some lenses alter the angle of light rays—concave, prismatic, polarizing, and bat-wing lenses. Concave lenses spread light broadly. Prismatic lenses use prisms to control light. Light sources need to be positioned correctly— the correct distance apart and the correct distance from the prismatic lens. Otherwise, the performance and appearance of the lens will be altered. In addition, tinted-prismatic lenses are available for fluorescent fixtures. Some can match tinted windows in smoke or bronze. Some are silver tints, either pale for aesthetics or deep for reducing reflected glare. Many choices are available. The choice should be dictated by design intent, comfort for the user, as well as aesthetics. (See *prismatic lens* in the Illustrated Glossary, Chapter 21.)

Polarizing lenses put light out vertically rather than both vertically and horizontally. As a result, direct and reflected glare are reduced. At the same time, contrasts, details and colors are improved. They are most effective when retrofitting a luminaire in the glare-producing zone (in front of the desk for pencil/paper tasks; behind the desk for computer monitor tasks). (See *polarized lens* in the Illustrated Glossary, Chapter 21.)

Bat-wing lenses broadly spread the most intense angles of light to the left and to the right into the glare-free zone (30 to 40 degrees), out of the veiling-reflection zone (0 to 30 degrees). Use bat-wing lenses when sources are to be mounted above and in front of the work surface, wherever it is.

Some lenses alter electronics—an electromagnetic shielding lens. (Fluorescent sources generate electromagnetic interference which is delivered with the light. The electromagnetism can affect equipment, such as radios, sound systems, medical devices, computers, TV, etc.) These lenses drain away the interference through grounding. (See *radio frequency interference* in the Illustrated Glossary, Chapter 21.)

Some lenses are impact proof. They are excellent for public spaces.

DIFFUSERS

Technically a diffuser is also a lens. But, diffusers scatter the light to soften it, whereas lenses redirect light rays to obtain a specific distribution. Diffusers are made of Lexan, glass, and acrylic. Opal glass is a true diffuser. It scatters but also absorbs, reducing the available light. Dropped opals provide the broadest distribution, including putting some light on the ceiling. Recessed drop opals do not.

Reflectors

Reflectors focus the light by their shape, position, and finish. Reflectors can be precise optical controls and focus light. Computers design precise, high-performance reflector shapes on the basis of ray traces. Space-age technology provides high-performance reflector finishes. Such reflectors control light rays and redirect them into the room rather than getting lost in

Cone reflector.

Ellipsoidal reflector.

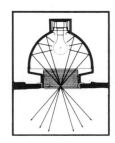

Reflectors and lens reprinted by permission of Capri.

the fixture. For example, with a high-performance reflector, a four-source fluorescent fixture can be a two-source fixture, saving electricity for both lighting and air-conditioning. Also, a high-performance reflector can redirect the light from a 70-watt halogen to equal that of a 300-watt non-halogen. For incandescent sources, an ellipsoidal reflector focuses the light to a point. It is precise and controllable. A parabolic reflector redirects light into parallel rays—very useful for work spaces.

Computer-designed reflector for high-angle pole light.
(See pg. 231 for candlepower distribution.)

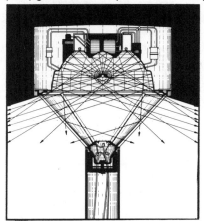

Reprinted by permission of Kim.

The position of the reflector to the source dictates the photometrics (the distribution of intensities of light from a luminaire)—wide, narrow, or asymmetric. If the source is incorrectly positioned, the distribution is disadvantaged and out of focus. Many well-designed incandescent fixtures allow positioning for different distributions. Some well-designed, single-source fluorescent fixtures also allow positioning. These minor adjustments can have major impacts. Adjustable fixtures allow the designer to have different light distributions and yet have all the fixtures look the same. Use them.

Reflector finishes can be specular, semispecular and matte. The finish dictates how the light rays will be redirected. Specular finishes can be space-age film at 96 percent reflective or can be mirrored aluminum at 85 to 87 percent. Specular finishes redirect light in the same, but opposite angle from which it is received. Semispecular finishes are alzak, an anodizing process like etching, or others, reflecting less and spreading a little. Semispecular and matte fin-

ishes control the light less than specular, but the source will not be mirrored on them. Matte-white paint reflects the least. Matte finishes diffuse light in all directions. Different applications require different reflectors.

Incandescent-cone reflectors have color choices that affect amount of light; gold reflects 90 percent of what it receives and black reflects only 70 percent. Likewise, cone reflector colors affect the color of the light. Use them for design advantages when appropriate.

Light rays cannot be condensed. They can only be redirected. Consequently, reflectors provide well-engineered surfaces for redirecting the light.

REFLECTANCES

Specular

Semispecular

Matte

power to reduce energy consumption

Businesses of all kinds utilize daylight to reduce their dependence on electricity. Photosensors can monitor the amount of daylight and adjust the electric lighting to provide constant illumination levels at desks or in shops. The payback time (the time required to save enough to equal installation costs) for incandescent lights is 1 to 2 years; and for fluorescent lights it is 2 to 5 years. In an incandescent system, photosensors create long-term savings with less electricity for lighting and air-conditioning and with less bulb replacements. In a fluorescent system, photosensors create savings with less electricity for lighting and air-conditioning only, and bulb life is not affected. Dimming incandescent light makes it more golden; dimming does not affect the color of fluorescent light, but it affects color rendering of surfaces. Photosensor systems operate automatically to dim or to raise light levels. They can be installed by owners or renters in new or existing structures.

Sample Electric Cost

A photosensor could save $484 in electricity at 6¢ per kilowatt-hour in an office lighted by daylight and 72 ceiling 40-watt fluorescent bulbs operating for 4,000 hours per year. With an installation cost of $1,775, the payback time would be 3.7 years.

A photosensor could save $678 in electricity at 6¢ per kilowatt-hour in a shop lighted by daylight and 33 ceiling 100-watt incandescent bulbs operating for 4,000 hours per year. With an installation cost of $800, the payback time would be 1.2 years.

Reprinted by permission from Lutron.

Additional energy reductions are possible because light source and ballast manufacturers have developed energy-conserving products. Some energy-conserving 65-watt incandescent sources equal the light of standard 100 watt. The 35-watt white sodium equals the light output of a 100-watt PAR incandescent and lasts 10 times as long. Since fluorescent sources are used in the majority of commercial spaces, reductions in fluorescent energy consumption are significant. First, determine how sensitive the tasks in the space will be to the level of illuminance.

In spaces that can accept lower levels:

- Replace every other source in a rapid-start fixture with an energy-saving source that balances the light output and reduces energy consumption for a choice of either 33 or 50 percent. All sources will appear equally bright.
- Install power reducers in-line with the ballast.
- Reduce source wattages to 34, 32, or 28 for 4 ft (1.2 m) fixtures.
- Specify reflectors.
- Retrofit polarizing lenses in fluorescent fixtures that now have prismatic lenses. Likewise, remove 2 of the 4 fluorescent sources per fixture.
- Some utility companies offer a rebate for energy-conservation measures, such as lens replacement, removal of sources, and high-power ballasts.

Sample Electric Cost

Retrofitting hall and stair ceiling- or wall-mounted fixtures with 9W compact fluorescent sources instead of 60W A incandescent.

$$\frac{48W \text{ saved} \times 4{,}000 \text{ hrs/year} \times 10c \text{ kilo. hr.}}{1{,}000} = \$19.20$$

In spaces where higher levels must be maintained:

- Use sources with higher efficacy (number of lumens produced for electricity consumed).
- Specify triphosphor fluorescent.
- Utilize energy-saving electronic ballasts, as opposed to core-coil ballasts and save 8 or more watts for each two-source, 40-watt luminaire.
- Replace lenses in fluorescent fixtures that put light in the glare-producing zone—60 to 90° above nadir—with lenses that put light into the 30 to 60° zone (bat-wing or polarizing). Task visibility will be enhanced.

Indeed, most of the energy conservation in electric lighting comes from the choice of light source. Sodium sources, as a class, conserve the most energy. Metal-halide is second; fluorescent, third, with mercury next and incandescent last. Of course, these classes overlap with some equalling the conservation of another class. Consequently, careful comparison of wattage and amount of light produced must be done before informed decisions can be made. But, these informed decisions are necessary. Lighting can be, for instance, 20 percent of the energy bill for retail stores. Benefits of reducing energy consumption should not be offset by loss of sales, as long as the visual and brightness impressions remain the same.

Sample Electric Costs

For accent lighting, substitute a 90W PAR/CAP/FL source for 150W PAR FL and have the same amount of light, but reduce the electric consumption.

$$\frac{60W \text{ saved} \times 4{,}000 \text{ hrs/year} \times 10c \text{ kilo. hr.}}{1{,}000} = \$24.00$$

Usually the color of light and efficacy are conversely related—that is, good color usually means only fair energy efficiency. However, this is not true for deluxe triphosphor sources, which have good color and high efficiency. Use them in commercial spaces requiring good color of light.

Likewise, in offices, benefits of reducing energy consumption should not be offset by loss of productivity, as long as the light for tasks comes from the best angle for visibility and yields illuminance needed for the age of the worker, the contrast of the task, and the speed and accuracy required.

Further, energy efficiency is important, because in all commercial installations, lighting is consumed for long hours and creates heat. Air-conditioning must keep internal temperatures in an acceptable range. (Heat from lighting can add 40 to 60 percent load on cooling commercial spaces.) Use energy-conserving lighting alternatives to reduce heat. In addition, in cooler climates, use daylight as an alternative to electric light. According to a Martin Marietta Energy Systems' study about reducing energy consumption, daylight created the greatest reduction in energy consumption, task/ambient light was second, and lighting system changes—fixtures, diffusers, and ballasts—was last.

Many utility companies and manufacturers, desiring any sort of energy conservation, offer a rebate for energy-efficient changes in existing structures or energy-efficient installation in new structures. They give rebates for efficient fixtures and/or sources. Check the current status of rebates in your area. Get two benefits, not just one.

Automatic Lighting Controls

The work of service businesses has people coming and going in and out of many spaces. Frequently, lights are left on unnecessarily for long periods. Controls can automatically switch lights on or off, as needed, thanks to microprocessor technology. Controls can dim lights when daylight is available and help reduce energy costs in commercial buildings by 20 to 50 percent. Controls are activated by time clocks or sensors. Sensors respond either to light or to people (their motion or their heat).

Time clocks are reliable to turn off lights. Time clocks are sophisticated. They can automatically adjust to seasonal time changes, can be manually changed, and can be staggered when turning on after electrical failure to prevent damaging power surges. Likewise, manual changes can be limited by a time clock timing the manual override.

Photosensors respond to light. They are reliable and also sophisticated. They adjust to the amount of light available. They can be preset for a specific amount, responding by dimming or by turning on/off. Likewise, photosensors can control several areas at preset levels; for example, a store sign at 100 footcandles (fc), display windows at 200 fc, and backwall security lighting at 10.

Ultrasonic sensors respond to motion. They are triggered (hopefully) by the right motion and the best ones remain triggered even though motion ceases during occupancy. Otherwise, the lights turn off, as they did at a middle school in Tallahassee. (The students loved the disruption; the teacher did not find it amusing.)

Infrared sensors respond to heat and are less likely than ultrasonic sensors to read inappropriate clues. But, the sensor must be carefully positioned to work correctly. Good positioning and careful calibration are essential for the success of these people-sensing controls. Infrared sensors work well for spaces up to about 800 sq ft (244 sq m). They are excellent security devices.

Controls should be selected to conform to the use of the space. A space with intermittent use does well with a people-sensing control. (A middle school does not have intermittent use.) A space with scheduled use does well with a time clock (a school or restaurant). A large space with windows along one or more walls does well with a photosensor responding to the daylight conditions (a high-rise, multiperson office).

Several controls can be combined into an energy-efficient system for small businesses. Large businesses need a more complex control system in a centralized location. All controls can have manual overrides for personal influence over lighting. Some controls receive signals through the electrical lines. Do not use them with high-intensity discharge (HID), because the HID ballast might short out the control. Controls are delicate devices.

Payback time varies for controls. It can be as little as nine months or as long as several years. It depends upon the size of the business, the length of time used, and the local electrical costs. Once paid back, payback becomes welcomed savings.

power to create atmosphere

Light has the ability to create atmosphere. The late John Flynn established categories of atmosphere or visual impressions due to light. One set of categories relates to perceptual impressions (for example, spacious versus confined). Another set relates to behavior-evoking impressions (relaxation versus tension). The third set relates to overall impressions (like versus dislike). These results plus continued testing at Penn State, have yielded some indications of light's influence on what people feel about their surroundings. For example, the subjective feelings of privacy are reinforced by nonuniform lighting with low amounts immediately near the occupant and higher amounts farther away. Further, it has been found that feelings of pleasantness can be reinforced by nonuniform lighting with emphasis on wall lighting.

Likewise, color has influence. For example, in our culture, blue has long been accepted as creating a cool atmosphere; red creating warmth. Consequently since light not only influences how we feel about a space but also how we perceive color, light and color together are powerful tools for creating atmosphere.

At the same time, the style of the fixture creates atmosphere. For example, chandeliers and pendants have been used as decorative statements for atmosphere, particularly in restaurants. They visually set a tone and carry out a decorative theme, particularly when the same style as the furniture or architecture in historic restorations. Sometimes the fixture style is different than the interior, creating a counterpoint, such as neoclassic furniture with a contemporary lighting fixture.

Further, the form of the light also can create atmosphere. For example, the form and color of neon sets a tone. Similarly, prismatic tubes, cold cathode, thin T-5 or T-6 fluorescent, fiber optics, low-voltage tube, tape, and string lights can affect the atmosphere. Visual liveliness is possible with these forms of lighting.

Moreover, framing projectors can create atmosphere with color and patterns. Projectors use incandescent or HID point sources and if desired, color filters. Further, adjustable shutters and a pattern (template) can shape the light beam. With a dimmer pack and control board, the patterns and colors can automatically or manually fade and/or crossover. The system creates a slow-moving painting of light that need not repeat for hours. With skilled preplanning, they are useful in store windows, showrooms, merchandise displays, night clubs, restaurants, and other places desiring a predetermined atmosphere. Their effect is more subtle than neon, fiber optics, and prismatic tubes.

Light Sources

TINTED LIGHT

Tinted light has the potential to create atmosphere in restaurants, hotels, and other public places for everyday use and for special events. Tints can be created within the source (gas or glass) or created external to the source (filters or coated bulb).

Color Filters

Color filters can be glass, gels, film, or plastic. They all change the color of the light and, therefore, the surface lighted. The more saturated the filter, the more saturated the color. Saturation of glass filters runs from 70 to 99 percent.

The higher the transmission percentage, the brighter the light will appear. Transmission of glass filters runs from 1 to 79 percent. Use the percent to compare brightness. Any color definable on the CIE's (International Commission on Lighting) 1931 Chromaticity Diagram can be duplicated in a glass filter.

Glass filters are dichroic, permitting a single

or a combination of colors to pass through at a particular wavelength and saturation. The filter rejects all other wavelengths.

Plastic sleeves fit over linear fluorescent sources for indoors or out. All sizes and colors are available.

All remote source lighting can have a color filter between the source and the light guide—fiber-optic or hollow prismatic film. The filter can be one color or a continuously variable color wheel for rainbow-like effects.

The amount of light transmitted through color filters varies. Yellow glass filters transmit the most—79 percent. Dark blue filters transmit the least—1 percent. Ultraviolet-blocking filters for fine art transmit only 70 percent. Thus, more candlepower might be needed to overcome this reduction.

Colored Sources

Colored sources can be cold cathode, fluorescent, laser, neon, PAR-38 dichroic incandescent, computerized light emitting diodes, (LED) and electroluminescent strips. What an array!

Cold cathode is linear, like fluorescent, and can be formed into gentle curves or follow any architectural configuration. It must be custom-made. Colors also are custom-made, for example subtle shades of pink, tangerine, peach, rose and apricot. The phosphors inside the glass create the color. The maximum length is 96 in. (2.4 m). It is surface mounted or hidden.

T-12 fluorescent sources are available in basic colors. They can be surface mounted, but the visual impact of the colors differs. A dimmer system can equalize the differences.

Laser light is one color. It can create visual excitement. It is created by a cascade of photons of light striking energized material and causing the release of more photons in a synchronized pattern. The wavelength is narrow. (Other light sources have broad wavelengths.) If the laser is an argon gas discharge type, the color is blue-green. If the gas is krypton, the color is red. A computer is needed for control and a projector to direct the light. Laser light is illusive and is usually seen in the air made visible by fogging. Laser light can be a powerful (and expensive) playtoy for the lighting designer.

Neon is linear source. Their color is created by either the gas in a clear tube or the combination of the gas and the glass. Tubes can be custom-made into any shape imaginable. Count on them for impact. Typically, they are surface mounted.

Dichroic incandescent sources selectively transmit only one color of light and capture the other colors. The dichroic color enhances surfaces of the same color. Sometimes, using two sources of different dichroic filtering creates a richer color. For example, a blue source and a green source imparts a rich aquamarine color on plant leaves. Dichroic light does not impart a tint to white or any other colored surface. The source should be hidden and only the enhanced surface seen.

Computerized light emitting diodes (LED) fixtures can create millions of colors from soft pastels to saturated hues and can do so with movement. A good combination! They are quiet and, like other LED's, use very little energy. Some of the fixtures fit tracks made by other manufacturers.

Electroluminescent systems create teal, green, yellow, crimson red, and medium blue light in widths up to 12 in. (30 cm) wide and 500 ft (152 m) long. They are surface mounted. They are stunning bands of soft color.

Designing with several tints in the same space is tricky. Tints do not appear equally bright even though they are same wattage. This phenomenon is particularly apparent when putting the same size cold-cathode or fluorescent sources together and finding the intensities so different. Dimmers can equalize. Experiment with any tinted light before using it. Colored light is an exciting variable. Create many marvelous visual impressions with it.

SMALL LOW-VOLTAGE LIGHTS

Small low-voltage sources can create atmosphere. They can be on tubes, tape, or string. They require transformers to reduce line voltage from 120 to 24 or 12 volts. Some have 4 circuits for chasing (blinking on and off).

Tube Lights

Tube lights in $\frac{1}{4}$- to $2\frac{1}{2}$-watt sources can create a twinkling, starlike atmosphere, depending on how closely they are positioned. These tiny sources

Low-voltage tube, tape, and string lights.

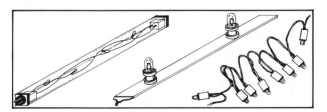

can be housed in tubes, flexible or rigid. The sources can be positioned 2 to 20 in. (5 to 51 cm) apart. The closer they are, the greater the sparkle. They can be hung as a curtain. The sources are distinctive and last for about 5 years. But, most are not replaceable and must be sent back to the factory for replacement.

Sample Electric Cost

In a large hotel lobby, a curtain of 9,000 tube lights costs 23¢ per hour to operate at 10¢ per kilowatt hour.

Tape or String Lights

Tape or string lights up to 5 watts can also create atmosphere. Further, the sources can be replaced individually when they burn out. They can be used to outline architectural edges. Some have color filter caps.

Low-voltage sources become guidelights when embedded in stair edges, handrails, and other architectural parts. They call attention to potentially dangerous changes. Also, they can be made into chandeliers of gigantic but delicate proportions gracefully raining into a space.

Likewise, small low-voltage sources can be encased in panels, like stars, making sparkling effects for walls and ceilings. String lights can be put in trees, threaded through shrubs and structures to create a twilight look of fireflies waiting to be caught.

Large Bare-Lamp Sources

Further, larger low-voltage sources in B, G, and S shapes with screw bases can create dotted lines of light with large-source identification for dramatic statements. Finally, with mirrors as backgrounds, all these sources become infinity lights reflecting and rereflecting for what seems forever. The only limit to using any of the atmosphere-producing sources is your clients' pocketbook. (Certainly not limited by your imagination.)

REMOTE SOURCE LIGHTING

Remote source lighting uses a single light source and transmits with total-internal-reflection light guides. The guides are either fiber optic or hollow prismatic. The total reflection is obtained by light rays entering a light guide at a specific angle and being re-reflected inside over and over again along the length of the guide. This phenomenon can be experienced. (See Indirect Remote Source Lighting in Chapter 9.)

Light guides use either an MR-16, a metal-halide, or a fusion source. Low-wattage (60), xenon-type metal-halide usually do not need to have a fan in the illuminator box. Filters or colored lenses can alter the light before it enters the guide.

Fiber Optic Light Guides

Fibers are either glass or plastic. They emit light from the ends (end-lit) or from the sides (side-lit).

Glass fibers transmit the light the farthest with the highest percentage of light. Glass has less color shift and lasts longer, but is more costly.

Plastic fibers can be thin or thick. They can become yellow and brittle over time. Core fibers are covered with light-confining cladding and an external protective sheathing.

End-lit fibers are either thin, from 1 to 10 mm (⅜ in. or less), or thick, 3 to 12 mm (less than ½ in.) and are bundled together. End-lit fibers can have a variety of fixtures with many lenses for different applications—accenting jewelry, downlighting artifacts in museums, signage, boat lighting, landscape floods, steplights, and short bollards.

Side-lit fibers are either round, oval, or square. The shape is dictated by the intended use—fitting between glass blocks or outlining architectural edges, swimming pools, and point-of-purchase displays.

Typically, fibers absorb 3 to 5 percent of the light per foot—more apparent with side-lit fibers. Thus, not quite all that is put in gets out. Fibers cannot make right-angle bends. Bend radius is critical to preserving the total internal reflection. Follow manufacturers' specifications.

A fiber light source must produce light at the acceptable half-angle, usually 30 to 40 degrees. Some sources do a better job than others. MR-16 spots or 150- to 250-watt halogen PAR sources can be used, but are not a highly efficient system.

The xenon metal-halide is much better. It was created to illuminate fibers. It is small (less than ½ in. or 12 mm) and gets light into fibers with less heat, usually eliminating the necessity for a cooling fan. At 60 watts, it produces 2,000 lumens for a 4,000-hour life, and restarts instantly, warming up to 50 percent light output in 20 seconds. (Other metal-halide sources require longer and need a fan for cooling.) Sunlight can be a no-energy fiber source!

Hollow Prismatic Light Guides

Prismatic light guides are hollow 3 to 8 in. (8 to

Glassblock walls can be lighted
with edge-lit fiber optics.

20 cm) round or square for side-lit, larger-scale applications. They have an illuminator with the light source, a hollow acrylic guide of prismatic film to transmit light, a diffusing film to block light where needed, and a mirror at the end of the guide (if a second light source is not required).

Ready-made guides are available for general illumination where access to changing sources is difficult or environments are sensitive and/or hazardous. Ready-made guides are 40 ft (12 m) long and two can be linked together. Guides can be custom-made for, among other things, imbedding in the underside of a handrail, incorporating into a cove around a room, and surfacing a luminous-panel desktop. Typically, metal-halide or fusion sources are used. Imagination, time, and money are the only impediments, to using remote sources.

Dimmers and Other Controls

Dimmers have the power to create atmosphere in commercial (and residential) spaces. Dimmers can fine tune a well-orchestrated lighting design. They can change lighting emphasis. For example, dimmers can change lighting on a display when one product is on sale. They can tone-up or tone-down the total lighting at a slow rate that is almost unnoticed by the people using the space. Toning down saves electricity, although it may be prompted by a time-of-day decision. For example, a restaurant can move from late-afternoon lighting to cocktail-hour lighting with a dimmer.

How does a dimmer work? An electronic (or solid-state) dimmer is basically a switch. It dims lights by turning the power on and off 120 times a second. Dimming occurs by controlling the amount of time the power is on versus the amount it is off. The longer the power is off, the more the light dims. When off, energy is saved and incandescent sources last longer.

Basically every light source is dimmable, but some sources can not be dimmed to dark. Some last longer if dimmed. Some change color. Some take up to 5 minutes to dim. In general, all incandescents can be totally dimmed and last longer if they are, but the color of the light becomes more red. Fluorescent can be dimmed to almost dark, do not last longer, and produce no color shift. Cold cathode and neon cannot be totally dimmed, do not last longer, and change color, depending upon their original color. (Experiment with them before specifying.) High-intensity discharge sources cannot be totally

dimmed, take up to 5 minutes to dim, do not last longer, and change color. Metal-halide and mercury shift to more blue-green; sodium shifts to more orange.

Different light sources require special dimmers and dimmable ballasts. Low-voltage incandescent need low-voltage dimmers. Low-voltage must be dimmed on the line-voltage side. Autotransformers dim low-voltage, but they are large and do not integrate well with complex dimming systems. For the most part, fluorescent, high-intensity discharge, and cold cathode require solid-state electronics for dimming. Metal-halide can be dimmed using a core-coil ballast and dimmer.

Unfortunately, dimming can create noise. Incandescent sources may buzz. To decrease the buzz, use sources with less wattage, use a smaller bulb size or use rough service type. If all else fails, get debuzzing coils. Never use a light dimmer to control a fan. But, a fan control can dim incandescent sources. When using a PAR and an electronic dimmer, the dimmer needs a debuzzing coil. When dimmed, fluorescent ballasts can hum and can vibrate if ballasts are attached to the fixtures. Solve the problem by purchasing an "A" rated or a solid-state ballast. If not, install the ballast remotely. In addition, dimming can cause flickering at the ends of the

How a dimmer works.

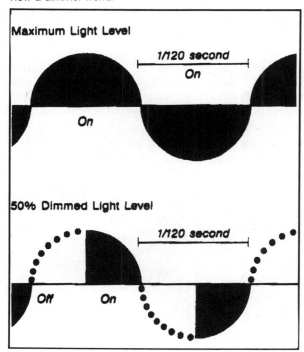

Reprinted by permission of Lutron.

fluorescent tube. Solve by only using a dimmer designed for the fluorescent source and by not dimming energy-saving fluorescent. Solid-state ballasts cure both noise and flicker problems. Do not mix a solid-state transformer and a solid-state dimmer, unless the dimmer is specifically designed for such a transformer.

If low-voltage halogen sources are dimmed and the bulb appears to darken, operate them at full power for 10 minutes in order to allow the tungsten cycle to regenerate. (Darkening does not affect the life of the bulb.) Do not operate dimmed low-voltage circuits without all the light sources in place and replace burned out sources promptly. Some low-voltage fixtures cannot be dimmed; check manufacturers' catalogs.

In general, dimmers were designed to dim, not to save electricity, unless in a dimming system responding to available daylight. Dimmers consume electricity themselves, even when off. The smallest dimmer consumes about ½ watt. Dimmers are not totally efficient. They spend 2 percent of their electricity producing heat. This heat is dissipated around the dimmer. Some newly engineered dimmers do not require a large wallbox to get rid of the heat. They are cool, slim and trim. Some dimmers can be controlled remotely, a convenience for quick changes in boardrooms, bedrooms, and banquet spaces.

On/off and dim control can be triggered by different kinds of inputs: light, time, movement, heat, and manual control. Controls that utilize light input are photosensors. They respond to light levels of both daylight and electric light. Controls that utilize time input are time clocks and load-shedding controls. Time clocks respond to a preset time. Load-shedding controls respond to utility companies' punitive electrical costs during peak electrical consumption times—the dinner hour. Controls that utilize movement and heat inputs are occupancy sensors. Manual controls respond to hands on the switch.

Each type of control has a reason for being chosen. Photosensors are excellent for switching to daylight. Time clocks are excellent for spaces used at specific times. Load-shedding controls are excellent for locations where utility companies have different utility rates at different times of the day. Occupancy sensors are excellent for turning lights on in spaces where people move in and out, particularly if too busy to turn on the lights. Manual controls are unsurpassed

for giving employees the feeling that they have some authority over their environment. But, lights might not be turned off again. Hence, time-clock controls can be added to manual controls and set time limits on additional light.

Each type of control has it own payback time. They are valuable for businesses that plan to stay in business for a couple of years when payback become savings for cashflow in a downturn or for covering costs of expansion.

Restaurant Lighting

Restaurants rely heavily upon light to create atmosphere. The pace of a restaurant is often set by the atmosphere created by the lighting. A brisk-paced atmosphere is created in fast-food restaurants by bright lights or lively forms of cold cathode, neon or other lighting forms. A leisurely pace is created by subdued lighting. Restaurants can utilize theater lighting applications for memorable interiors. Create an atmosphere according to the type, the standard of accommodation, and the potential trade of a

Pools of light next to darkness create an intimate atmosphere.

restaurant. Although atmospheres may differ, lighting for essentials does not. The essentials are tables, serving counters, work stations, and public facilities. Light is needed at these places. Lighting design for essentials can be different. Needless to say, the more time and money available, the more opportunities there are for designing powerful lighting beyond the essentials. Nonetheless, essential lighting can be designed effectively.

Restaurant Lighting Guidelines

- Large sources of light (large luminous ceilings, ceiling panels, or large fixtures) need to yield soft, not harsh light.
- Low-level lighting slows down the eating pace.
- Fixtures hanging below the ceiling (chandeliers or pendants) should relate to tables or other furniture. Conversely, fixtures on the ceiling can relate to the space or its divisions.
- Intense downlights should not be over customers' heads. Harsh shadows are created under eyes and noses.
- Tabletop candles alone do not yield enough light for a whole restaurant.
- Pools of light on tabletops next to darkness create an intimate atmosphere.
- Direct light (downlight from pendants or recessed fixtures) on tabletops creates sparkle on the tabletop objects.
- Bare, low-wattage bulbs (in chandeliers, surface-mounted ceiling fixtures, wall fixtures, or flexible tubes) create sparkle. None should be over 15 watts each.
- Wall-hung fixtures add to the decoration but do not produce sufficient light for most restaurants.
- Wall-washing emphasizes the wall finish and also reflects light.
- If low illumination is the design goal, then higher illumination should be available for cleaning the restaurant.
- Programmable dimmers can change lighting levels for different atmospheres without complicated settings; for example, moderate level for a banquet, brighter with a focus on the after-dinner speaker, and subdued for dancing.

DESIGN CONSIDERATIONS

Consider the time of day a restaurant will be used. If it is to be used during the daytime and has no windows, the illumination level must be higher than at night. The contrast of light between the inside and outside should not be too great—one to five times is considered a reasonable ratio. Consequently, during the day when the outdoor illumination level is high, the interior level must also be relatively high; otherwise

Large sources of light need to be soft.

eye adaptation could be momentarily painful. (Remember the last time you came out of an afternoon movie on a sunny day and how much your eyes hurt for a moment or two?) Photosensors are available to automatically control the light inside a restaurant on the basis of the level outside and to maintain a comfortable contrast.

Consider the various uses of the restaurant space. For example, it may be rented for birthday parties, club banquets, business meetings, and wedding receptions. Each function requires a different atmosphere. Preset or programmable dimmers can create whatever atmosphere is required with no guesswork. Once programmed, they operate with a simple on-off switch.

Consider the volume of the space and the effect desired when choosing fixtures. Normally, larger volumes of space accept larger visible fixtures. Smaller volumes accept small fixtures. However, hidden fixtures (recessed, cornice, cove, valance, luminous ceiling or panel) are good for both small and large spaces. Combine the style of fixtures with the design of the interior, arrangement of the space, and

Wall-washing reflects light.

structural constraints for the desired effects. Stylized fixtures can create an atmosphere themselves. Overall, visible fixtures should be pleasing to look at (lighted or unlighted), mechanically operable, easily maintained, and long-lasting. The range of effects are broad and hard to describe. In fact, professional illustrators hate to draw nighttime restaurant interiors because the atmosphere created is so difficult to reproduce graphically. For lighting designers, the process of creating atmosphere cannot be prescribed. Suggestions throughout this book will assist in designing restaurant lighting.

In designing, be sure to consider the contrast between the kitchen and the dining space. If the light in the dining space is subdued, make sure that bright kitchen light does not blast into the dining area. Otherwise, the artfully created atmosphere for the customers will be destroyed. Likewise, it will be too much contrast for waiters' and waitresses' eyes.

Places of Worship

Lighting reinforces the atmosphere created in places of worship (churches and temples). The atmosphere varies from sanctuary for meditation, shrine for ceremonies, and hall for social gatherings. The lighting atmosphere must also change. Preset dimmers are invaluable for this job. Symbolism in places of worship must be reinforced by lighting. Some structures have a grand scale, with soaring architecture. Others have a more intimate scale embracing the congregation. Some are pretentious; others plain. Whether awesome or humble, lighting can be designed to help evoke these feelings.

Often the lighting comes after the congregation is unhappy with the lighting originally installed and the lighting job is a retrofit. Consequently, low-voltage has the advantage. It can be carried on 12-volt wires, easy to hide and acceptable by local codes, in most cases. In addition, edge-lighted fiber optics, prismatic tubes, and prismatic film in tubes are also useful. Further, since historic restoration has had an upsurge, many manufacturers specialize in custom fixtures—a boon for lighting places of worship. With many sources, fixtures, and distributions to choose from, the job is usually a matter of convincing the building committee that the design is right.

In addition to atmosphere, places of worship have the same requirements as conference rooms for light on the speaker's face and reading matter, walking area, vertical displays, and congregation's reading matter. (See Conference Rooms in this chapter.) Likewise, requirements for reading music must be met. (See Playing a Musical Instrument, Chapter 13.) Low-voltage PAR sources can throw light a long way for such tasks; don't ignore tasks.

power for steering

Light has the power to steer people in the direction you want them to go. Bright enough light attracts their attention and permits them to recognize what they are looking for—the exit, the public facilities, the cashier station. Make the areas you want to be seen three times brighter than the surrounding area and reduce questions about directions.

Neon and Cold-Cathode Lighting

For many years, neon has been used as outdoor signs to steer people to places of business. It is unexcelled in attention getting and ability to be molded into any shape. It is useful for accenting interior architectural features and enhancing commercial spaces. Likewise, cold cathode is also excellent. (Cold cathode is a relative of fluorescent, which is hot cathode.) These two sources can be molded. Consequently, they have gained popularity in residential spaces as artful sculpture. In commercial spaces they can outline interior architecture, illuminate handrails and coves, and can be luminous walls and sculpture. However, they can be overpowering, noisy and too functional-looking. Skill is required to use such lighting.

Fiber Optics

Fiber optics can produce bright light at the end or along the edge of the fiber. End-lit fibers are used to create patterns of light to attract attention and influence people. The patterns are controlled by computers. The color is created by a color wheel. Special decorative effects like starfields, fireworks, and flower arrangements can be created. The light for both end- and edge-lit

Cold cathode light emphasizes the architecture with brilliance.

fibers is created by a bright point source—low-volt MR-16 or a laser. Edge-lit fibers are used like neon, to create lines of light. They can outline architectural edges (stairs, roofs, etc.), spell out words, form abstract designs, or illuminate a surface. The fibers can be made of glass (hard to use) or plastic (flexible or rigid and easy to cut). They can be bundled together. The light comes out the end in a 30-degree cone, but does not travel very far. Fiber optics give the designer a wide palette to create exciting lighting.

Emergency Lighting

In business spaces, emergency lighting is required for a minimum of 90 minutes to comply with codes. Some emergency fixtures are designed to use auxiliary electrical power; some use batteries. Some are functional-looking units; some are architecturally integrated. Some are simply energy-producing packs to fit into the existing fixtures. Choose the most aesthetic type to suit your pocketbook. Be careful where the emergency lighting points. Emergency lighting can confuse people, especially if it glares in their eyes. It should be designed to get people to the exits in fire or dense smoke. Rising smoke obscures emergency lights and exit signs near ceilings. But, emergency light sources near the floor are visible. Battery pack units are available that plug into a wall duplex outlet. Installation is easy and the lens spreads the light broadly. In addition, flashing directional arrows at floor level leading the way out are excellent.

vocabulary for professionals

21

The language of lighting is confusing. Sometimes two terms mean the same thing—illumination and illuminance. Sometimes a term means two different things—skylight, meaning a window in the roof or light from the sky. Further, measurements based on feet are slowly being changed to different terms based on meters. The communication process is likewise impeded because of regional differences in building terms. For example, a recessed electrical connection can be called a junction box, an electric box, a j-box, a tangle box, or a splice box. In addition, many different people design lighting: lighting consultants, architects, interior designers, landscape architects, electrical engineers, manufacturers' reps, electricians, and lighting showroom staff. These people were not trained in the same way and do not talk the same language. However, professionals should speak the language of the lighting profession. But, the professional language often conflicts with the language used by the general public. For example, a professional says "lamp" and means what the general public calls a "light bulb." On the other hand, the general public says "lamp" and means what a professional calls a "portable luminaire." Therefore, lamp manufacturers make light bulbs, unless, of course, they are making portable luminaires. What a mess! For clarity without confusion, this book refers to light bulbs or lamps as "light sources," which they are, but cautions readers to use the word "lamp" for light source when talking to other professionals.

Professional Terms	Everyday Terms
Accent lighting	Spot lighting
Ambient light	General illumination
Controls	Switches, dimmers
Fenestration	Windows
Fenestration on the roof	Skylight
Ground light	Light reflected from the ground
Hard wiring	Built-in wiring
Housing	Part of the fixture surrounding the bulb
Illuminance	Light (meaning that which is produced)
Illuminance	Illumination
Lamp	Light bulb
Light source	Light bulb
Linear source	Fluorescent or tubular incandescent
Luminaire	Fixture and light source

Professional Terms	Everyday Terms
Luminaire	Light (meaning an object producing light)
Nonuniform lighting	Different amounts in the space
PAR lamp	Floodlight
Point source	Incandescent, mercury, sodium, metal-halide
Portable luminaire	Table or floor lamp
Reflected and direct glare	Glare
Retrofitting	Putting a different light source in an existing fixture
Skylight	Light reflected from sky and clouds
Solar radiation	Sunlight
Task light	Worklight
Uniform lighting	Same amount of light throughout a space

illustrated glossary

Lighting terminology is highly technical. Definitions are precise, qualifying, and scientific. Most defy memorization. Researchers who develop and test theories about lighting need such definitions. Designers who plan lighting can utilize the technology better with less rigorous definitions. Therefore, this illustrated glossary intends to clarify professional definitions without, hopefully, severely sacrificing technical quality.

The terms are divided into measurement, lighting fixture, light source, and lighting system and power supply categories. Some terms belong to two categories with different meanings, such as, controls. Light fixtures have controls (baffles, etc.) and lighting systems have controls (dimmers, etc.). Most terms, however, fall into one category. Use these definitions to develop basic understanding.

MEASUREMENT TERMS

Aiming Angle: Angle from which the light falls on the surface lighted. (In this book, 0° aiming angle is perpendicular to the surface being lighted.)

Beam Spread: Angle of light from a specific source on one plane within which the amount is 50% of the maximum. Beam-spread angle does not show asymmetrical light distribution. A quantity measurement.

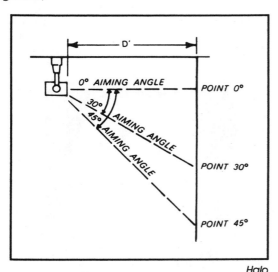

Halo

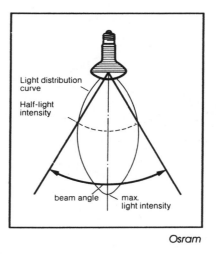

Osram

Angle of Incidence: Angle from which the light strikes a surface. A quantity and quality measurement.

Angle of Reflection: Angle from which the light leaves a surface. Always equal to the angle of incidence. A quantity and quality measurement.

Brightness Unit (BU): Relative brightness of different light sources or lighted surfaces. A quantity and quality measurement.

Calculators: Hand-held programmable calculators are excellent for computing zonal-cavity method of calculation,

Candlepower (cp): The intensity of light radiated in a particular direction by a light source. It is measured in candelas. Candelas divided by the distance squared (unless adjusted by the sine or the cosine of the angle of light if not 0°). A quantity measure.

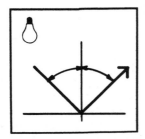

$$\text{footcandles} = \frac{\text{candelas}}{\text{distance}^2}$$

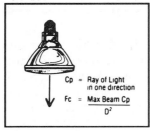

Candlepower

Cp = Ray of Light in one direction

$Fc = \dfrac{\text{Max Beam } Cp}{D^2}$

Sylvania

Candlepower-Distribution Curve: A polar plot showing the intensity of light in a given direction. Particularly good for determining the angle of maximum light intensity. The plot indicates candelas from a luminaire or a source located in the center. Two curves are plotted if beam pattern is asymmetric, if two different sources, or if both vertical and horizontal distributions. A quantity measurement.

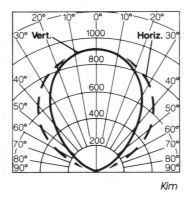

Kim

CapCal: Trade name for a measurement system to test luminance within a visual scene. The measure uses both video and computer technology to capture, store, retrieve, and analyze luminous information, as the eye would. Output is the expected level of visual performance, the brightness contrast of the task, and the overall brightness level. Quantity and quality measurements.

Chromaticity: Descriptive term identifying where the color of the light would be plotted on the CIE standard color diagram. A quality measurement.

CIE (International Commission on Lighting): An international organization on the science, technology, and art of lighting for developing standards and procedures.

Coefficient of Utilization (CU): A measure of the efficiency of a luminaire delivering light to a horizontal work plane, taking into account luminaire efficiency and light reflectances within the space. A quantity measurement.

Color Preference Index (CPI): Measure of how well a source renders colors flatteringly or most pleasingly. Sometimes called flattery index. A quality measurement with the human factor in it.

Color Rendering Index (CRI): A measure of color shift of surface colors from what was measured when illuminated by a standard source. (The CIE standard sources are: tungsten lamp, mean noon sunlight, and average daylight.) Incandescent (tungsten) light and north-sky light are both rated 100. The index ranges from 1 to 100. The higher the number, the "truer" the color of objects appear. Only compare CRI's of sources with similar color temperatures (+ or −300K). Otherwise, the information is misleading. (For example, cool-white fluorescent has a 66 CRI, deluxe cool-white 89. Both are 4200K and can be compared.) A quality measurement.

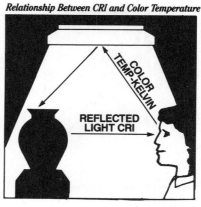

Relationship Between CRI and Color Temperature

Sylvania

Computers: Computers can handle large batches of numbers to give sophisticated printouts of expected illuminance on vertical or horizontal surfaces. The level of sophistication depends upon the hardware (computer) capacity and software (program) capabilities. Computers with graphics capacity can utilize one- and two-point perspective graphics software to show light. These graphics permit the professional to visually judge the lighted environment before installation.

Contrast Ratio: The relationship between the luminance of an object and its immediate background, such as contrast between letters and screen background on a computer monitor.

Contrast Rendition Factor (CRF): Apparent contrast of a task and its background under various lighting conditions. A quality measurement.

Correlated Color Temperature (CCT): The color of light expressed as a temperature that an incandescent object would have to reach when heated to produce light of the same color. Lower temperatures are reddish; higher temperatures are bluish-white. Temperature is in Kelvin degrees (K). The range for man-made sources is from 2000 to 7500K. A quality measurement.

Efficacy: Number of lumens produced for electricity consumed. A quantity measurement.

Exitance: Reflected light (used to be foot-lamberts).

Footcandle (fc): Measure of light produced by one candle uniformly onto a surface 1 ft² in area from 1 ft away. Translate footcandles into lux approximately by multiplying footcandles times 10. A quantity measurement.

$$\frac{footcandles}{ft^2} = lumens$$

$$\frac{lux}{meters^2} = lumens$$

IESNA (Illuminating Engineering Society of North America): A North American organization for the science, technology, and art of lighting to develop and publish standards and procedures.

Illuminance: Measurable light falling on a surface (illumination). Unit is footcandle or lux. A quantity measurement.

Initial Footcandles: Amount of light expected from the light source when new. A quantity measurement.

Intensity: Magnitude of light from a source or a luminaire.

Inverse Square Law: The law describing the phenomenon that the amount of light decreases by the square of the distance it travels. A quantity measurement.

Isochart:

Color temperature with color rendering index in parentheses.

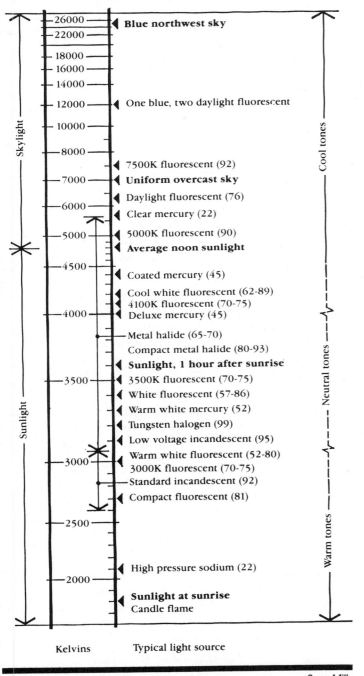

Kelvins	Typical light source

Sam Mills

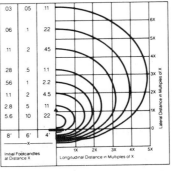

Kim

Isochart: An isochart shows the footprint or beam pattern of light on a surface. Units are either in footcandles or candelas. When the beam pattern is symmetrical in all four quadrants, one quadrant is shown. When the pattern is symmetrical in only two quadrants, two are shown.

K: Kelvin degrees (See *Correlated Color Temperature*.)

Light Effectiveness Factor (LEF): Percentage resulting from dividing the glare-free light in a space by the total light. Values range from 10 to 200%. Spaces with LEF around 100% are judged more positively than spaces with lower values. A quantity and quality measurement.

Light Meter: Meter to measure luminance exitance of about one degree in the visual scene (small). A quantity measurement.

Light Reflectance Percentage (LR): Light reflectance percentage or average amount of light reflected from a particular color surface. Also called reflectivity. A quantity measurement.

Lumen: Unit of luminous flux equal to the light emitted in a solid angle by a point source of 1 candela intensity.

Luminance: Measurable light reflected from an illuminated surface. A quantity measurement. (The subjective appraisal of luminance is brightness.)

Luminance Exitance: Total amount of light redistributed by a lighted surface. The measure is lumens per sq ft or meter. A quantity measurement.

Luminance Ratio: The contrast between the brightest and the darkest reflected light of surfaces.

Luminous Flux: Light emitted by a source. Measurement unit is lumen. A quantity measurement.

Luminous Intensity: Light produced by a source in a given direction.

Lux: Metric measure of light produced by one candle uniformly onto a surface 1 meter² in area from 1 m away. A quantity measurement.

Maintenance Factor: Used in zonal cavity calculations to obtain average, not initial, levels of light. It takes into account depreciation losses over time, such as losses due to dirt on room surfaces (RSDD), on luminaires (LDD), and lamp-lumen losses (LLD) and burnouts. The factor ranges from .5 to 1. Multiply all the contributing factors together or consult the interpretive charts for luminaire types (direct, indirect, etc.), room cavities and time. A quantity measurement.

Metameric Match: When two colors match under one light source but do not match under another (the interior designer's nightmare).

Nadir: Zero angle of light aimed straight down to the floor. A quantity measurement.

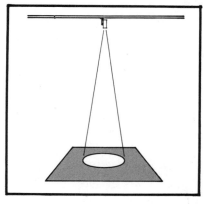

Lightolier

Photometrics: Defines function and performance of a luminaire. Used to compare and evaluate lighting system alternatives. Data usually contains: candlepower distribution, shielding angle, efficiency, spacing to mounting-height ratio and coefficient of utilization. Quantity measurements.

Point-by-Point Calculations: Calculation method to determine the illumination on a surface from any angle. Used for nonuniform lighting.

Radiosity: Calculation of diffusion from matte surfaces or of transmission of light through translucent surfaces or through particles in the air.

Ray Tracing: Calculations of reflections of light rays from surfaces.

Reflectance: Measure of reflected light from a surface. Reflectance equals the amount of light reflected from the surface divided by the amount of light falling on the surface. Measurement is a percentage. A quantity measurement.

$$\text{reflectance} = \frac{\text{lumens coming off surface}}{\text{lumens falling on surface}}$$

Reflectance

Relative Visual Performance: Measure of performance of a visual task taking into consideration contrast, speed and accuracy, and age of viewer. A quantity and quality measurement.

Room Cavity Ratio: Ratio of room height to width expressing how well the light can be utilized. Ratio ranges from 1 to 10. Used in zonal calculations. Ratios of 1 are large rooms, have high ceilings or are long and narrow; ratios of 10 are small rooms. A quantity measurement.

$$RCR = \frac{5 \times H\,(L+W)}{L \times W}$$

Spacing Criteria (SC): Formula for determining how far apart to install luminaires for uniform lighting, based on mounting height.

$$MH \times SC = D$$

Spectral Distribution Curve: Curve plotted from the amount of energy in each wavelength for a particular light source. A quality measurement.

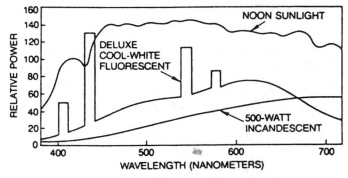

Ontario Hydro

Task Visibility: Task size, contrast and brightness. These all affect the amount of light needed to perform the task satisfactorily. A quantity measurement.

Visual Comfort Probability (VCP): Evaluation of direct glare, judged by people seated at the worst possible location for glare in a room with a lighting system using a specific fixture. The rating is the percentage of people who describe the system as comfortable. (The higher the VCP, the better.)

Visibility Unit (VU): Relative clarity of light or ability to see clearly. A quality measurement.

Watts: Unit of electrical power. A quantity measurement.

Zonal Cavity or Lumen Calculation: Calculation method to determine number of light fixtures needed to provide a specific light level or vice versa. Room configuration, surface reflectance and light-loss-over-time are considered. Used for uniform lighting. A quantity measurement.

LIGHTING FIXTURE TERMS

Adjustable: Fixture that permits positioning the light source to aim it vertically and/or horizontally. Some adjust more than others. Good for sloped ceilings, accenting, and/or wall-washing.

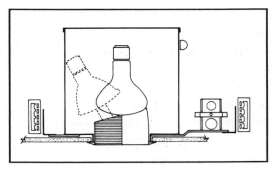

Capri

Air Handling: Fixture that also supplies heated and cooled air in and out of the space.

ANSI: American National Standards Institute develops specifications for ballasts and proper lamp performance.

Baffle: (Two kinds—horizontal and vertical) Horizontal baffle is a series of rings on sides of the aperture that reduce brightness. A vertical

Horizontal baffle.

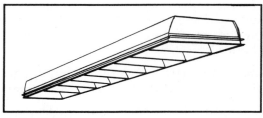

Vertical baffle.   *Metalux*

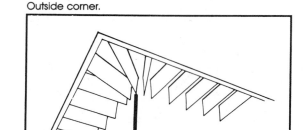

Outside corner.

Columbia

baffle is a series of parallel fins that block the view of the source and brightness of light above 45 degrees.

Ballast: Regulator to control voltage, current and wave form for fluorescent and HID sources. It must match source type, wattage, and voltage.

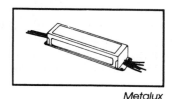

Metalux

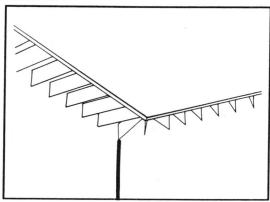

Inside corner.

Columbia

Ballast Power Factor: Low or high, depending upon how much power used. High power factor ballasts use less electricity. Many local codes require them.

CBM: Certified Ballast Manufacturers label indicates that ballast meets ANSI standards.

Channel: Metal housing to hold a linear fluorescent tube. Can be mounted end-to-end (butt) or overlapped (staggered).

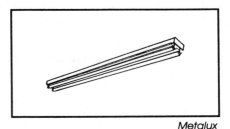

Metalux

Controls: (Two kinds—fixture and light system) Fixture controls are reflectors, baffles, shutters, and other fixture parts to redirect or cut off light. (See *controls* under Light System and Power Supply Terms.)

Corners: For fluorescent perimeter lighting, either an inside corner or an outside corner.

CSA: Canadian Standards Association label indicates that ballast meets CSA standards.

Cut-Off Angle: The angle up from 0 degrees nadir where the light source can no longer be seen. (Cut-off angle plus shielding angle equal 90 degrees.)

Diffuser: Glass or plastic aperture cover that transmits and scatters light rays.

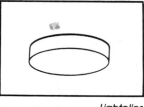

Lightolier

Downlight: Recessed or surface-mounted fixture that aims the light down.

Efficiency: Measure of the lumens from the light source that emerge from the fixture. (A bare bulb has the highest efficiency, but efficiency does not take glare into account.)

ERL: Environmental Research Laboratories, one of the independent groups that tests photometrics.

ETL: Electrical Testing Laboratories, one of the independent groups that tests photometrics.

Fiber Optics: Fibers of glass or plastic that carry light. Fibers can be rigid or flexible. Some fibers carry light internally from the source to the end of the fibers (end-lit). The light is brilliant, but goes only a short distance beyond the end. Some fibers uniformly emit light along the fiber itself (edge-lit). Both are used for decorative lighting. Edge-lit fibers can be used for ambient lighting.

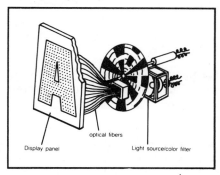
Fiber optics (end-lit).

Display panel   optical fibers   Light source/color filter

Lazarus

Filter: Glass, gel, or coating that alters the color or distribution of light.

Framing Projector: Fixture providing control of light by shutters, iris, or metal template cut in a pattern.

Fresnel Lens:

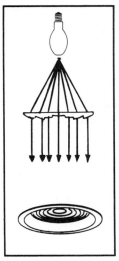

Sylvania and Lightolier

Fresnel Lens: Lens that redirects light rays to become parallel, functioning like a convex lens without the thickness.

High- or Low-Bay Luminaires: For HID sources in industrial spaces, particularly warehouses.

Housing: Metalwork that surrounds the bulb in recessed fixtures. Some housings with thermal guards can be covered with insulation and still be safe (IC type). Some with thermal guards must be kept 3 in. (7.6 cm) away from insulation (T type). (A thermal guard deactivates the fixture if heat builds up.) Some do not have thermal guards and are for suspended ceilings (suspended-ceiling type).

ITL: Independent Testing Lab, one of the independent groups that tests photometrics.

Lamp Image: Reflection of the lighted bulb on the fixture's cone.

Lamping and Aiming: Installing light sources; aiming and adjusting fixtures for best distribution of light.

Lenses: Fresnel, spread, opal, polarized, prismatic, convex, and concave for altering the delivery of light.

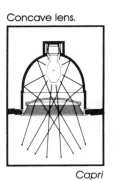
Concave lens.

Capri

Light Pipe: Trade name for a prismatic tube or box-shaped pipe that uniformly transmits light through the sides for the whole length. Light sources can be mounted remotely at the end of the pipe since the efficiency is so high. Excellent for difficult or dramatic applications.

Light Tapes (or Strips): Rigid or flexible tapes with bare low-voltage sources of 1 to 10 watts.

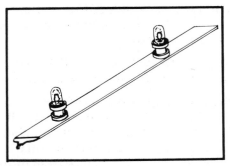
Light tape.

Excellent for creating decorative light or for small places.

Light Tube: Rigid or flexible plastic tubing with ¼- to 2-watt low-voltage sources. Excellent for calling attention to steps and other architectural features.

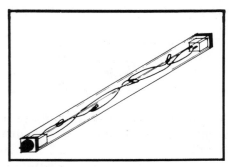
Light tube.

Louvers: Vertical, cube cell, small- or deep-cell parabolic, barndoors, and other attachments for fixtures to alter the shielding angle and light distribution.

Deepcell parabolic louver.

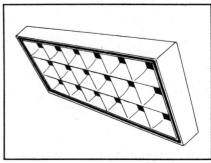
Metalux

Barndoors

Lightolier

Low-Profile Downlight: Recessed fixture that is designed for shallow ceiling-joist depth (plenum).

Luminaire: Lighting fixture with a light source.

Luminaire Efficiency: Lamp lumens emitted by a luminaire divided by lamp lumens generated by the light source.

LSI: Lighting Sciences Inc., one of the independent groups that tests photometrics.

National Electrical Code (NEC): Codes of established national requirements for electrical safety to prevent fires from electrical arcing and excessive heat build-up in fixtures.

National Electrical Manufacturers Association (NEMA): Association that has developed a classification of beam spreads for floodlighting. The classification ranges from 10° spread, type 1 to 130°+ spread, type 7. Asymmetric beam floodlights have 2 designations, indicating horizontal and vertical spreads.

Open-Reflector Fluorescent: Bare fluorescent tubes with reflectors and no lenses, often used in large retail stores. They produce inescapable direct glare.

Prewired Recessed Fixture: Conduit, connector, and junction box attached. (Saves installation time, but prohibited by some local building codes.)

Polarized Lens: Lens that alters the glare-producing characteristics of light rays—the radiation in all directions—into a more usable vertical direction.

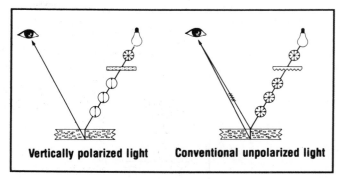
Vertically polarized light   **Conventional unpolarized light**

Polarized Int'l

Prismatic Film: Film that transmits light while letting it uniformly escape through the film, unless blocked by opaque film. Used inside tubes or box-shape pipes for edge- or end-lighting. Light source is positioned at one end and can be remotely mounted.

Prismatic Lens: Lens that uses the nonparallel sides of a prism to redirect the light rays.

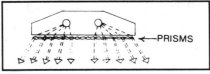

Lightolier

Radio Frequency Interference: Radio signal that can be produced by ballasts and dimmers which interferes with sound systems and sensitive electronic equipment. Special lenses block such interference.

Radio-frequency shield lens.

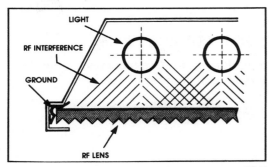

K-S-H

Reflectors: Cone, ellipsoidal, multiplier, parabolic, wall-washer, downlight/wall-washer and others. Some are created by computer ray-trace technology. All alter the delivery of light.

Parabolic reflector.

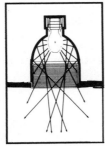

Capri

Shielding Angle: Angle down from a line at the bottom of fixture's edge where the light source can first be seen. (Shielding angle plus cut-off angle equal 90 degrees.)

Shielding angle.

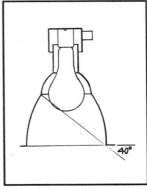

Edison Price

Slope Adaptor: Adaptation device for compensating for sloped ceilings when installing recessed fixtures.

Spacing Criteria (SC): Formerly called spacing to mounting height ratio. The ratio of the distance between fixtures (on center) to the mounting height above the surface to be lighted. Used for maximum fixture spacing in uniform lighting. Acceptable uniformity is minimum and maximum light levels not more than $1/6$ above or below the average level. Wide distribution fixtures have a criteria of 1.0 or more; medium have .7 to .9; and narrow have .6 or less.

Transformer: Device that changes voltage. For example, low-voltage fixture must have a transformer to reduce voltage. Transformers are either magnetic or electronic and can be in the fixture or remotely mounted. (Electronic are smaller.)

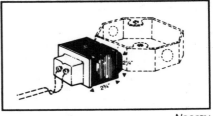

Neoray

Trim: Baffles, cones, rims, and other treatments for apertures of downlights.

Types of Fixtures: Direct, semi-direct, general diffused, direct-indirect, semi-indirect, and indirect. These types were derived from the candlepower distribution curves.

Direct

Semi-direct

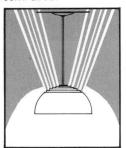

Direct-indirect

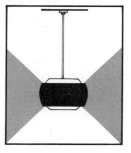

General diffused

Semi-indirect

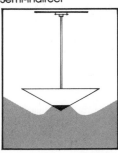

Indirect

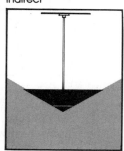

UL: Underwriters Laboratories, a non-profit organization that tests, certifies, and labels approval that luminaire conforms with national safety standards required by most local codes. Only use fixtures that have UL labels and have custom fixtures manufactured to UL standards.

LIGHT SOURCE TERMS

Abbreviations for Ordering from Suppliers: For incandescent sources, a notation of 25A17/RS reads:

25	watts
A	shape
17	size (divide by 8 for diameter inches)
RS	rough service

$$\frac{17}{8} = 2\frac{1}{8} \text{ in. diameter}$$

For fluorescent sources, a notation of F96T12/CW reads:

F	Fluorescent
96	in. long
T	tubular shaped
12	size (1½ in. diameter)
CW	cool white

For HID sources, a notation of M400/C/BU-HOR reads:

M	Metal-Halide
400	watts
C	phosphor coated
BU-HOR	base can be positioned horizontally

These abbreviations are in the manufacturers' lamp (bulb) catalogs.

Ambient Temperature: Temperature surrounding the light source. Fluorescent sources are affected by temperatures surrounding them. The energy-saving versions require above 60°F; others require above 50. Special ballasts are needed for colder conditions. Light output is also affected with higher or lower than 77°F.

Arc Tube: Light-emitting part of HID sources.

Bases:

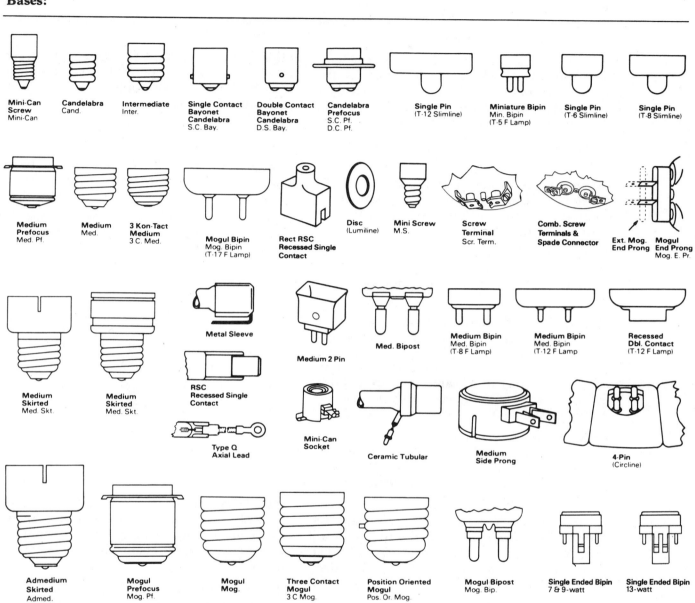

Sylvania

Bulb Finishes: X means deluxe and best color of light for the source type. Others are:

    A = amber
  AIC = amber iridescent clear
    B = blue
    G = green
   IC = iridescent clear
   IF = inside frost
    R = red
  SKP = soft pink
    W = white
    Y = yellow

Bulb Size: Determine by dividing the number indicated by 8 to obtain the bulb's diameter in inches.

$$A19 = {}^{19}/_8 \text{ or } 2{}^{3}/_8 \text{ in.}$$

Dichroic: PAR source with a special filter (color-selective or heat rejecting) that permits only one color or little heat to be transmitted. With one color, surfaces of the same color are greatly enhanced. With heat rejecting, merchandise is not depreciated. Useful in retail display lighting.

ER: Ellipsoidal reflector, incandescent source with an internal reflector that has a focal point 2 in. (5.08 cm) in front of the bulb and does not trap light inside a deep recessed fixture.

Filament: Straight or coiled wire for incandescent sources. The light-emitting part of the source. The smaller the filament relative to light output, the more precise the control of the light. (Halogen sources have the smallest.)

Flood: Incandescent or HID source either PAR or R with a wide beam of light.

Fluorescent Strobing: Turning on and off 120 times per second. Strobing causes visual discomfort for some people. Electronic ballasts help reduce this effect.

High-Pressure Sodium: High-intensity discharge source which is very energy efficient and available in color-improved versions for interiors.

K-Beam Classification: Method of analyzing the principle parts of a beam from both source and fixture for predicting visual impact in accent lighting. (This method was developed because two beams at the same angle from different spot sources do not necessarily produce the same visual edge.)

Lamp Depreciation: Lumens produced by a light source decrease over time. Consequently, the amount produced at mid-life of the lamp is less than the initial amount. Manufacturers indicate this amount as "mean" lumens.

Lamp Lumens: Total light output. Manufacturers specify lumens expected initially for all types of sources. For fluorescent, they also list lumens at middle of source's life. For HID, they list mean lumens.

Sylvania

Laser Light: Light produced by oscillations of atoms or molecules. A thin beam of light produced in one wavelength or color. Of the several types of laser devices, the most common are gas discharge and optically pumped. The light can be damaging to the eyes. A license is required for its use.

Light Center Length (LCL): Distance from center of light-producing part (filament or arc) and bottom of base. Important for correct positioning with reflectors.

Linear Source: Fluorescent or tubular incandescent.

Luminous Efficiency: Lumens produced by the light source divided by watts required for the lamp.

Maximum Overall Length (MOL): Distance from top of bulb to bottom of base. Important for proper fit in fixtures.

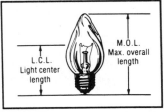

Sylvania

PAR: Parabolic-aluminized-reflector source made of pressed glass with a built-in reflector, like an automobile headlight. Yields more, but harder-edged light beam than reflector (R) sources.

Phosphor Coating: Coating on the inside of a tube or bulb that produces light when bombarded by ultraviolet light, which is produced within the source. It is the light-emitting part of fluorescent and some HID sources.

Point Source: Light-producing part of a source that is almost a point. Hence, a point source can be precisely focused to direct the light. Halogen is the best. Other incandescent and clear HID are almost as good. Phosphor-coated HID and compact fluorescent are not point sources.

R: Reflector source made of blown glass with a built-in parabolic reflector. Yields a softer and somewhat more indistinct-edged light beam than a PAR source.

Shapes:

INCANDESCENT LAMPS

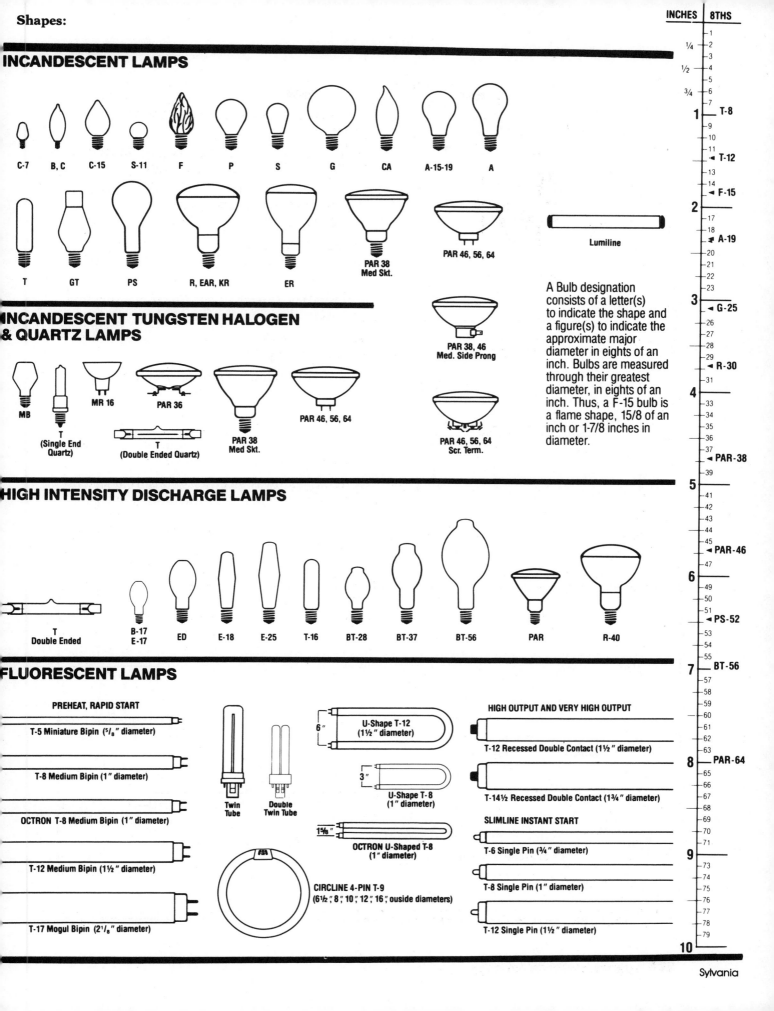

C-7 B, C C-15 S-11 F P S G CA A-15-19 A

T GT PS R, EAR, KR ER PAR 38 Med Skt. PAR 46, 56, 64 Lumiline

INCANDESCENT TUNGSTEN HALOGEN & QUARTZ LAMPS

MB T (Single End Quartz) MR 16 PAR 36 T (Double Ended Quartz) PAR 38 Med Skt. PAR 46, 56, 64

PAR 38, 46 Med. Side Prong

PAR 46, 56, 64 Scr. Term.

A Bulb designation consists of a letter(s) to indicate the shape and a figure(s) to indicate the approximate major diameter in eights of an inch. Bulbs are measured through their greatest diameter, in eights of an inch. Thus, a F-15 bulb is a flame shape, 15/8 of an inch or 1-7/8 inches in diameter.

HIGH INTENSITY DISCHARGE LAMPS

T Double Ended B-17 E-17 ED E-18 E-25 T-16 BT-28 BT-37 BT-56 PAR R-40

FLUORESCENT LAMPS

PREHEAT, RAPID START

T-5 Miniature Bipin (5/8" diameter)

T-8 Medium Bipin (1" diameter)

OCTRON T-8 Medium Bipin (1" diameter)

T-12 Medium Bipin (1½" diameter)

T-17 Mogul Bipin (2¹/₈" diameter)

Twin Tube Double Twin Tube

CIRCLINE 4-PIN T-9
(6½", 8", 10", 12", 16" ouside diameters)

6" U-Shape T-12 (1½" diameter)

3" U-Shape T-8 (1" diameter)

1⅝" OCTRON U-Shaped T-8 (1" diameter)

HIGH OUTPUT AND VERY HIGH OUTPUT

T-12 Recessed Double Contact (1½" diameter)

T-14½ Recessed Double Contact (1¾" diameter)

SLIMLINE INSTANT START

T-6 Single Pin (¾" diameter)

T-8 Single Pin (1" diameter)

T-12 Single Pin (1½" diameter)

Scale markings:
¼ 1 2 3
½ 4 5
¾ 6 7
1 — T-8
9 10 11 — T-12
12 13 14 — F-15
2 15 16 17 18 — A-19
19 20 21 22 23
3 — G-25
25 26 27 28 29 — R-30
31
4 32 33 34 35 36 37 — PAR-38
39
5 40 41 42 43 44 45 — PAR-46
47
6 48 49 50 51 — PS-52
53 54 55
7 56 — BT-56
57 58 59 60 61 62 63
8 — PAR-64
65 66 67 68 69 70 71
9 72 73 74 75 76 77 78 79
10

Silvered Bowl: A-shaped source with silver on the bowl, radiating the light back into a broad reflector on the fixture. (Commonly used in Europe; not so common in the U.S.)

Halo

Spill Light:

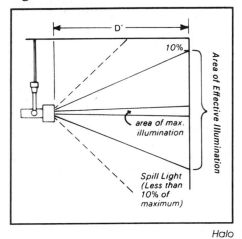

Halo

Spill Light: Spill light is outside the intense beam area (50% of maximum). It has 10% of the candlepower. Some sources have wider spill areas than others.

Spot: Incandescent or HID source, either PAR or R, with a confined beam of light.

LIGHTING SYSTEM AND POWER SUPPLY TERMS

Accent Light: Light for highlighting surfaces of objects, architecture, landscape and interior furnishings.

Alternating Current (ac): Electricity that reverses its flow 60 times a second.

Ambient Lighting: Light for the space in general (as opposed to task light).

Ampere (A): Units of flow of electricity.

$$\text{Ampere} = \text{watts} \times \text{volts}$$

Ceilings: Types for fluorescent installations are: suspended (exposed grid, concealed tee, screw slot, Z-spine, or metal pan) and nonsuspended (plaster or plasterboard).

Circuit: Pathway electricity runs from generator (utility company) to structure, through light source and back to generator. Branch circuits are within the structure. Circuits are opened and closed by switches.

Electricity delivery compared to water delivery.

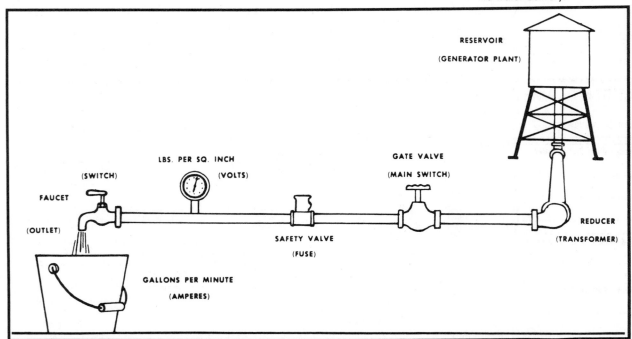

Florida Power

Circuit Breakers or Fuses: Safety valve of the electrical system turning off or blowing when the amps are overloaded. Overloading would cause heat resulting in fire.

Controls: (Two kinds—light system and fixture) Light system controls are dimmers, timers, and sensors to reduce or turn light off and on, automatically, preprogrammed, or by hand. (See *Controls* under Lighting Fixture Terms.)

Direct Current (dc): Electricity that flows in one direction. A battery is direct current.

Direct Glare: Glare from an overly bright light source in the field of view. When excessive, it is called disabling glare.

Lightolier

Facade Lighting: Floodlighting the exterior of a structure for security or for creating an image.

Ground Fault Interrupter: Protective device that shuts off the electricity at the least little change of current, preventing electrical shocks for electrical appliance users.

Group Relamping: Relamping all light sources at a predetermined time whether burned out or not. Relamping when about 70% of the sources are expected to be burned out. In large installations, this system saves labor costs and keeps light quantity as high as possible.

Initial Cost: Cost of fixtures and light sources.

Kilowatts (kW): 1000 watts. A kilowatt hour (kWh) equals 1000 watts used for 1 hour. Utility bills read in kilowatts. Determine electricity cost by:

$$\frac{W \times \text{hrs. used} \times \text{days}}{1000} \times \text{cost kWh} = \text{cost}$$

Lighting Energy Budget: Lighting performance requirements in various states limit the amount of electricity usable in specified spaces.

Low-Voltage Lighting: Fixtures and light sources that operate at 6, 12, or 24 volts, not the standard 120 or higher volts. Requires a transformer.

Offending Zone: Area on the ceiling from which a direct-light source would cause veiling reflections on a task. For glossy reading or pencil and paper tasks, the zone is in front of the task surface (desk) and at an angle of incidence between 0 (nadir) and 30°.

Operating Costs:

Annual operating costs—

$$\frac{kW}{hrs.} \times \frac{\#\,hrs.}{used} \times \frac{\text{cost kW}}{hr.} = \text{operating cost}$$

Cost to replace burned out sources—

$$\text{Av. replacements/yr.} \times \frac{\text{source}}{\text{cost}} + \frac{\text{labor}}{\text{cost}} = \frac{\text{cost to}}{\text{replace}}$$

Payback—

$$\frac{\text{Cost of lighting system}}{\text{annual energy} + \text{lamp replacement savings}}$$

Reflected Glare: Glare that is reflected from a surface, either architectural (marble wall), interior (glass-top table) or task (shiny magazines).

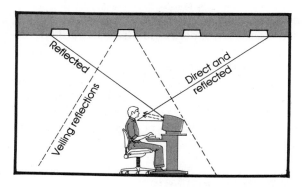

Return on Investment: Savings on utility costs, relamping expenses, and increased income or productivity due to light balanced against initial, operating, and maintenance costs of the lighting system.

Task Lighting: Light for visual work, usually localized to a small area.

Veiling Reflection: Reflected image of a light source or a brightly lighted surface that obscures the visual task.

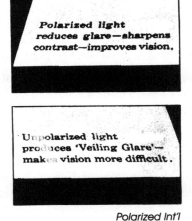

Polarized light reduces glare—sharpens contrast—improves vision.

Unpolarized light produces 'Veiling Glare'— makes vision more difficult.

Polarized Int'l

Voltage (Volt): Unit of the force of the electricity (like water pressure). Light sources are manufactured for specific voltage ranges. A higher operating range burns out the light source quicker. A lower operating range allows incandescent sources to last longer, but delivers less light. Standard voltage is 120 with a 115 to 125 range for sources. Higher voltage, 240, has a 230 to 250 range. Voltages go up to 277 and 480.

Voltampere (VA): watts

$$volts \times ampere = watts$$

Watts (W): Amount of electricity consumed by the light source and ballast or transformer, if there is one.

$$watts = volts \times amperes$$

Wall-Washing: Illuminating a vertical surface.

calculation methods

To provide sufficient amount of light for whatever is being designed (attracting people's attention, permitting work to be accomplished, or balancing other lighting, etc.), professionals need to determine beforehand how to do it. The footcandles needed can be determined. Then, the required wattages, the number and position of fixtures can be calculated. The light created should enhance visibility, not destroy it. Enhancement is not contained in footcandle determination alone. Enhancement is the combination of position and angle of light delivery, of task size and contrast, and of reflectances in the space. Many calculation methods are available to help with these determinations: interpreted calculations and hand- or computer-generated calculations.

Interpreted Calculations

Interpreted calculations are suitable for some installations, but not all. Information to be interpreted is published by manufacturers. It is based on photometric data compiled by independent testing labs. The information can be found in better catalogs. Unfortunately, between catalogs, conformity of terminology, similarity of material and measurement are lacking. For example, some measurements are in candelas; some in footcandles. Professionals are on their own to make the information equal and put to accurate use. (Footcandles equal maximum beam candlepower divided by distance squared.) Lots of traps exist.

Interpreted calculations come in many forms: light distribution, spacing criteria, and coefficient-of-utilization data. Each form is different and based on different measurements. Some are section views of light distribution. Some are footprints or plan views. Some consider room reflectances; some do not. Some consider room configurations; most do not. All intend to predict the intensity of light falling on a surface. (None intend to predict the visual impact, which is, of course, affected by a lot of factors, including the visual acuity of the observer.) All show initial footcandles, not the average over time. Nonetheless, they are helpful in making design decisions, especially in the beginning.

Calculation graphics are reprinted by permission of manufacturers noted.

INTERPRETED CALCULATIONS

| KIND | INFORMATION GIVEN | | | | INFORMATION TAKEN INTO ACCOUNT | | USE |
	Light Distribution	Light Intensity	Area of Light	Distance from source	Inter-reflectances	Room Configurations	
Candle-power distribution curve	section view	yes	yes	no	no	no	point-by-point calcula-tions, quick judgements
Beam spreads	section view	yes	yes	yes	no	no	accent, quick judgements
Lighting-performance data	plan view	yes	yes	yes	no	no	wall-wash
Isocharts	plan view	yes	yes	yes	no	no	accent, wall-wash
Footcandle distribution	plan view	yes	yes	yes	yes	no	wall-wash
Spacing criteria	assumes uniform lighting	no	no	must be known	no	no	how far apart for fixtures
Coefficient of utilization tables	no, assumes uniform lighting	no	no	assumes 30" above floor	yes	yes	zonal cavity calcu-lations
Quick calcu-lation charts	no, assumes uniform lighting	no	yes	assumes 30" above floor	yes	yes	quick comparisons

LIGHT DISTRIBUTION

Light distributions are section or plan views of lighting. Section and plan information are graphs and sometimes backed up with numerical tables.

Candlepower-Distribution Curves

Candlepower-Distribution Curves are section views. They are cross-section graphs showing where and how the light is distributed from a source or luminaire. They show light intensity

Candlepower curve and light distribution.

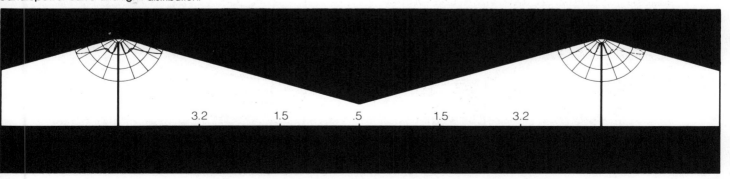

3.2   1.5   .5   1.5   3.2

Kim

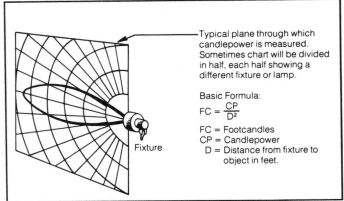

Typical plane through which candlepower is measured. Sometimes chart will be divided in half, each half showing a different fixture or lamp.

Basic Formula:

$$FC = \frac{CP}{D^2}$$

FC = Footcandles
CP = Candlepower
D = Distance from fixture to object in feet.

Fixture

Candlepower distribution curve.   *Kim*

at various angles. The 50-percent-of-maximum angle is particularly important. It is the basis for beam-spread information. Likewise, maximum-intensity angles show the concentration of light. Two curves are drawn if the light distribution is not symmetric, if two different sources can be used in the same fixture, if the aim could be either vertical or horizontal, or if the fixture is adjustable.

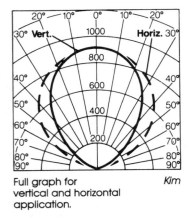

Full graph for vertical and horizontal application.   *Kim*

Use Candlepower-Distribution Curves for judging effectiveness of a particular source or luminaire for the distribution and intensity of light desired. However, do not be fooled. The scales of the graphs are different. When comparing source to source (or luminaire to luminaire),

Compare intensities of same source and different fixtures.

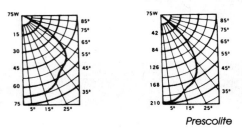

Prescolite

observe the scale. A quick visual comparison, even within one manufacturer's catalog, could be misleading. Compare the intensity scales (on the left of the chart); both are 75-watt source luminaires.

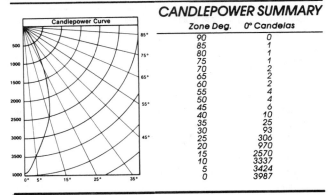

CANDLEPOWER SUMMARY

Zone Deg.	0° Candelas
90	0
85	1
80	1
75	1
70	2
65	2
60	2
55	4
50	4
45	6
40	10
35	25
30	93
25	306
20	970
15	2570
10	3337
5	3424
0	3987

Capri

In the 75-watt example above, the distribution is narrow. The luminaire would be suitable for a single-source accent to illuminate a small surface or as multiple, closely spaced sources for grazing a textured wall. The chart shows the maximum angle of light is around 8 degrees.

½ graph for asymmetric wall-wash downlight.

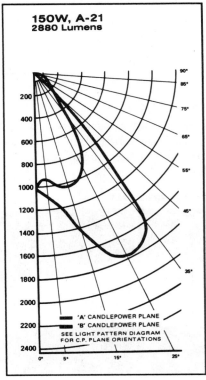

150W, A-21
2880 Lumens

'A' CANDLEPOWER PLANE
'B' CANDLEPOWER PLANE
SEE LIGHT PATTERN DIAGRAM
FOR C.P. PLANE ORIENTATIONS

Prescolite

Legend

Outer beam spreads are to 10% maximum candlepower
Inner beam spreads are to 50% maximum candlepower
(Beam spreads are indicated)

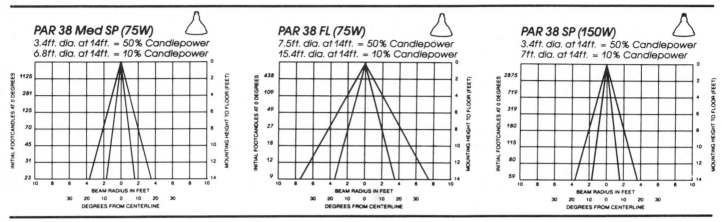

Beam spread.   *Capri*

Numerical tables of Candlepower Summaries or Candela Values are often published along with candlepower-distribution curves. They are useful for obtaining candela values when figuring point-by-point calculations by hand or computer.

Beam Spreads

Other section-view information is Beam Spreads. Beam Spreads show width and intensity at various distances from the source. The versions are called: beam-spread graphs, photometry charts, lamp- or lighting-performance data, lamp-selection guide, and cone-of-light charts. By whatever name, the information is similar, but not always the same. Graphs and charts usually show profiles of light distribution. From this information, initial footcandles, mounting height, and beam width can be read. Use these graphs and charts as far as they go.

In general, beam spreads are good for quick calculations indoors or out. They are particularly good for designing accent lighting. If the beam pattern is asymmetrical, two spreads are shown. If the beam pattern is symmetrical, both 50 and 10 percent of maximum spreads are shown. The 50-percent area will be the effective visual beam. The 10-percent area will be the last really visible light. For example, if the 50-percent area is 4 ft (1.2 m) wide and the 10-percent area is 8 ft (2.4 m) wide, everything within the 4 ft will be highlighted and nothing beyond 8 ft will be appreciably lighted. The vis-ible edge of light is the line where the light drops to 10 percent. Any light beyond 10 is called spill light. Use the lowest watt source with a 50-percent beam spread that covers just the surface to be highlighted at the level desired. The result will be good lighting and energy savings.

Lighting-Performance Data

Tables of Lamp- or Lighting-Performance Data are very useful for designing special effects indoors or out. They show distance, footcandles, and beam length/width at various aiming angles. (The beam width/length is the plan view or the footprint of the light.) The tables allow the user to take the known and determine the unknown. If the known information is: size of surface to be lighted, ceiling height, footcandle desired and aiming angle, then the unknown: the light source and position of luminaire can be determined. For example, at 30° on a 3 ft (.9 m) by 1½ ft (.45 m) vertical surface with 30 fc intended, the position of the luminaire would need to be 4½ ft (.75 m) from the wall and the light source would need to be 36W PAR36 NSP.

On the other hand, if the known information is: position of fixture (probably already installed), aiming angle, and light source (probably the one that fits the fixture), then the unknown: the beam size and footcandles can be determined. For example, a track light mounted 3 ft (.9 m) from the wall at the same aiming angle, with the same source, can provide 71 fc on a 2 by 1 ft (.6 by .3 m) area for displays.

PAR36 Lamps	Narrow						
	25W PAR36 (5.5V)	25W PAR36 VNSP	25W PAR36 NSP	35W PAR36 SP	36W PAR36 VNSP (T-H)	36W PAR36 NSP (T-H)	50W PAR36 VNSP
FOOTCANDLES (On Beam Center At 6')	833	458	117	50	556	142	561
BEAM SPREAD (To 50% Max. CP)	1.5°	2.5°×5°	8°×10°	14°×20°	5°	10°	4°×7°
MAX. CANDLEPOWER (Candelas)	30,000	16,500	4,200	1,800	20,000	5,100	20,200
RATED LIFE (Hours)	1,000	2,000	2,000	2,000	2,500	2,500	2,000
COLOR TEMPERATURE	2,600°K	2,550°K	2,550°K	2,550°K	3,000°K	3,000°K	2,550°K

0° AIMING ANGLE — Illumination on Horizontal Plane

	D	FC	L	W	D	FC	L	W	D	FC	L	W	D	FC	L	W	D	FC	L	W	D	FC	L	W	D	FC	L	W
	10'	300	0.3'	0.3'	8'	257	0.3'	0.7'	6'	117	0.8'	1.0'	4'	113	1.0'	1.4'	10'	200	0.9'	0.9'	6'	142	1.0'	1.0'	10'	202	0.7'	1.2'
	15'	133	0.4'	0.4'	12'	114	0.5'	1.0'	8'	66	1.1'	1.4'	6'	50	1.5'	2.1'	15'	89	1.3'	1.3'	8'	80	1.4'	1.4'	15'	90	1.0'	1.8'
	20'	75	0.5'	0.5'	16'	65	0.7'	1.4'	10'	42	1.4'	1.7'	8'	28	2.0'	2.8'	20'	50	1.7'	1.7'	10'	51	1.7'	1.7'	20'	50	1.4'	2.4'
	25'	48	0.7'	0.7'	20'	42	0.9'	1.7'	12'	29	1.7'	2.1'	10'	18	2.5'	3.5'	25'	32	2.2'	2.2'	12'	35	2.1'	2.1'	25'	32	1.7'	3.1'

30° AIMING ANGLE — Illumination on Horizontal Plane

	D	FC	L	W	D	FC	L	W	D	FC	L	W	D	FC	L	W	D	FC	L	W	D	FC	L	W	D	FC	L	W
	8'	304	0.3'	0.2'	6'	298	0.3'	0.6'	5'	109	0.9'	1.0'	3'	130	1.0'	1.2'	8'	203	0.9'	0.8'	5'	133	1.2'	1.0'	8'	205	0.7'	1.1'
	12'	135	0.4'	0.4'	10'	107	0.6'	1.0'	7'	56	1.3'	1.4'	5'	47	1.6'	2.0'	12'	90	1.4'	1.2'	7'	68	1.6'	1.4'	12'	91	1.1'	1.7'
	16'	76	0.6'	0.5'	14'	55	0.8'	1.4'	9'	34	1.7'	1.8'	7'	24	2.3'	2.9'	16'	51	1.9'	1.6'	9'	41	2.1'	1.8'	16'	51	1.5'	2.3'
	20'	49	0.7'	0.6'	18'	33	1.0'	1.8'	11'	23	2.1'	2.2'	9'	14	3.0'	3.7'	20'	32	2.3'	2.0'	11'	27	2.6'	2.2'	20'	33	1.9'	2.8'

30° AIMING ANGLE — Illumination on Vertical Plane

	D	FC	L	W	D	FC	L	W	D	FC	L	W	D	FC	L	W	D	FC	L	W	D	FC	L	W	D	FC	L	W
	4'	234	0.4'	0.2'	3'	229	0.5'	0.5'	2'	131	1.1'	0.7'	1'	225	1.0'	0.7'	4'	156	1.4'	0.7'	2'	159	1.4'	0.7'	4'	158	1.1'	1.0'
	6'	104	0.6'	0.3'	4'	129	0.7'	0.7'	3'	58	1.7'	1.0'	2'	56	2.1'	1.4'	6'	69	2.1'	1.0'	3'	71	2.1'	1.0'	6'	70	1.7'	1.5'
	8'	59	0.8'	0.4'	5'	83	0.9'	0.9'	4'	33	2.3'	1.4'	3'	25	3.1'	2.1'	8'	39	2.8'	1.4'	4'	40	2.9'	1.4'	8'	39	2.2'	2.0'
	10'	38	1.0'	0.5'	6'	57	1.0'	1.0'	5'	21	2.8'	1.7'	4'	14	4.1'	2.8'	10'	25	3.5'	1.7'	5'	26	3.6'	1.7'	10'	25	2.8'	2.4'

Lighting-performance data.

Lightolier

Read manufacturers' directions carefully. Their information has different starting points. Some assume that the surface being lighted is the work surface. Therefore, the surface is calculated at 30 in. (.76 m) above the floor. Quick calculations to light the floor could not be determined from such data. Likewise, some assume that the zero aiming angle is directly down and others assume that it is perpendicular to the surface lighted. Either way, the aiming angles of 30° and 60° are reversed. Consequently, 30° is 60° in the other system (45° stays the same in both).

The difference can throw interpretations off dramatically. Check basis of data carefully.

Be aware that no interreflectances are included in these tables. If other light sources are in the space, the final intensity might be higher.

Isocharts

Plan views of light distribution are Isocharts. The charts show the footprint or beam pattern of light on a surface. Isocharts are primarily used for outdoor lighting. Outside, surfaces are surrounded by darkness. Hence, the beam pattern

Compare isofootcandle chart with candlepower distribution chart for same luminaire.

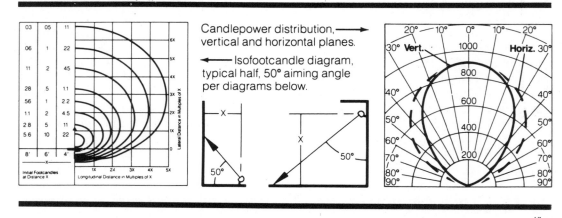

Candlepower distribution, vertical and horizontal planes.

Isofootcandle diagram, typical half, 50° aiming angle per diagrams below.

Kim

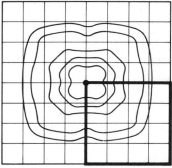

¼ chart for symmetrical distribution.

Kim

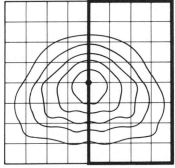

½ chart for asymmetrical distribution.

Kim

is highly visible. An Isochart provides intensity levels at positions on the surface from a single luminaire in footcandles or candelas. Be aware that no interreflectances are included. Outdoors, interreflections are usually negligible. (Indoors they are often significant.)

The grid indicates units of mounting heights (1MH, 2MH, etc.) and the isobars are the various intensity levels for different mounting heights, keyed to the numerical tables accompanying the chart. If the beam is symmetric, only ¼ of the chart is shown; if asymmetric, ½ is shown.

Like most interpreted information, the charts do not always contain the specifics needed for the design job. Consequently,

calculations by other methods are recommended. Be aware that installation measured output is never exactly like calculated output.

Footcandle Distribution
Footcandle-Distribution information is also plan views. Many versions exist. Some are wall-washing footcandle distributions. They are numerical tables showing light levels below and between luminaires. Use them for illuminance levels from luminaires at specific distances back from the wall and at specific spacing between. For example, with luminaires 3 ft (.9 m) back from the wall and 3 ft (.9 m) apart, the illuminance level 3 ft down the lighted wall would be 37 footcandles directly below the fixture and

Footcandle distribution data.

MANUFACTURER: CAPRI LIGHTING
CATALOG NUMBER: *MV23-T025F*
DESCRIPTION: *R-LAMP WALL WASH*
LAMP: *175W R40 MERC MED, 5700 LUMENS*
TEST REPORT NUMBER: *LRL 283-3B1*

Single wall wash/downlight 3 feet from wall.
Initial vertical illumination on wall on 1" x 1" grid.

Distance from ceiling in feet.	0.0	1.0	2.0	3.0	4.0	5.0	6.0	7.0	8.0
			Distance on wall in feet — To right of fixture centerline. Illuminance—Footcandles						
1.0	11.2	9.03	5.25	2.88	1.43	0.76	0.44	0.26	0.15
2.0	21.1	17.3	11.4	7.05	3.24	1.57	0.79	0.42	0.25
3.0	19.6	17.5	11.6	8.17	4.08	2.19	1.24	0.72	0.42
4.0	16.4	15.2	11.1	7.86	4.31	2.55	1.38	0.83	0.53
5.0	16.9	15.2	9.71	6.16	4.93	2.62	1.59	0.97	0.56
6.0	16.2	14.8	9.88	5.67	4.04	3.09	1.81	1.04	0.66
7.0	13.5	12.9	9.57	5.85	3.61	2.70	2.05	1.29	0.69
8.0	10.7	10.3	8.42	5.73	3.64	2.39	1.86	1.42	0.88
9.0	8.48	8.21	6.87	5.30	3.55	2.33	1.64	1.32	0.94
10.0	6.78	6.58	5.60	4.48	3.41	2.25	1.54	1.17	0.89

Table represents half of illuminated area.

Array of wall wash/downlights on 3 ft. ctrs 3 ft. from wall.

Distance from ceiling in feet.	0.0	1.0	2.0	3.0
		Illuminance—Footcandles		
1.0	17.4	16.5	16.5	17.4
2.0	35.9	33.5	33.5	35.9
3.0	37.2	35.3	35.3	37.2
4.0	33.5	33.1	33.1	33.5
5.0	30.8	32.4	32.4	30.8
6.0	29.4	31.8	31.8	29.4
7.0	27.3	28.8	28.8	27.3
8.0	24.0	24.8	24.8	24.0
9.0	20.7	20.9	20.9	20.7
10.0	17.3	17.8	17.8	17.3

Array of wall wash/downlights on 4 ft. ctrs 3 ft. from wall.

Distance from ceiling in feet.	0.0	1.0	2.0	3.0	4.0
		Illuminance—Footcandles			
1.0	14.2	12.9	11.4	12.9	14.2
2.0	27.8	26.4	24.3	26.4	27.8
3.0	28.2	28.5	25.7	28.5	28.2
4.0	25.6	26.4	24.9	26.4	25.6
5.0	27.3	24.9	22.6	24.9	27.3
6.0	25.0	24.6	23.4	24.6	25.0
7.0	21.4	22.8	23.2	22.8	21.4
8.0	18.8	19.8	20.6	19.9	18.8
9.0	16.5	17.2	17.0	17.2	16.5
10.0	14.5	14.5	14.3	14.5	14.5

For Gold Reflector use .9 Multiplier Factor

Capri

about 35 footcandles in between.

Footcandle-Distribution information is used for accent lighting, particularly with adjustable fixtures, both recessed and surface-mounted. Be aware that interreflectances within the space are taken into account in this data.

SPACING CRITERIA

Interpreted information also is in the form of Spacing Criteria (SC). This information has been called spacing ratio (SR) or spacing to mounting height (S/MH). By any name, the information is the same. Use it for uniform lighting. Wide-distribution fixtures have a criteria of 1 or more. Medium-distribution fixtures have .7 to .9. Narrow-distribution has .6 or less. The criteria is multiplied by the mounting height to determine the distance apart the luminaires can be installed to get uniform lighting. For example, a 10 ft (73 m) height with luminaires of a 1.4 criteria, needs to have luminaires placed 14 ft (102.3 m) apart.

$$10 \text{ ft} \times 1.4 = 14 \text{ ft}$$

Rule of Thumb for Distance Apart with Uniform Lighting
Mounting height times spacing criteria equals distance apart.

$$MH \times SC = \text{distance apart}$$

The criteria does not tell what level the uniform lighting will be. Nor does it take into consideration interreflectances. Consequently, the spacing might be too close. On the other hand, these ratios do assure that the variability will not be greater than ⅙ of the average light level. This variability is suitable, except for visually critical tasks. For example, without additions from interreflectances (or subtraction of dark surface colors), if the luminaire provides 50 footcandles, the variability of light would be from 42 to 58 footcandles.

$$\frac{1}{6} = .166 \qquad .166 \times 50 \text{ fc} = 8.3 \text{ fc}$$
$$50 + 8.3 = 58.3 \text{ fc}$$
$$50 - 8.3 = 41.7 \text{ fc}$$

COEFFICIENT OF UTILIZATION

Interpreted calculations are also in the form of Coefficient-of-Utilization information. Coeffi-cient of Utilization differs from other data because interreflectances and room configurations are taken into consideration. Therefore, they represent the total amount of light that would fall on the lighted surface.

Sometimes the information is published right out of photometric reports as numeric tables with room-cavity ratios on one axis and surface-reflectance percentages on the other. They show the Coefficient-of-Utilization factor. (This factor is used in the zonal cavity formula for determining footcandle level that can be expected or for the number of luminaires required when lighting uniformly.)

Room Cavity Ratio	Ceiling Cavity Reflectance								
	80			50			10		
	Wall Reflectance								
	50	30	10	50	30	10	50	30	10
	Coefficients of Utilization for Fl. Cav. Refl. = 20%								
1	56	55	54	53	52	51	49	49	48
2	52	51	49	50	49	47	45	44	43
3	49	47	45	47	46	44	45	44	43
4	46	44	42	45	43	41	43	41	40
5	44	41	39	42	40	38	41	39	38
6	41	38	36	40	37	36	38	37	35
7	38	35	33	37	35	33	36	34	33
8	36	33	31	35	32	31	34	32	30
9	33	30	28	32	30	28	32	29	28
10	29	26	24	29	26	24	28	26	24

These coefficients were computed by the zonal-cavity method, I.E.S recommended practice, and prepared from the candlepower distribution data in photometric report LRL 285-2C

Coefficient-of-utilization table. *Capri*

Other times, the information is published as charts. The charts are called "Quick Calculations." The charts show initial footcandles, number of fixtures or fixture spacing, and some kind of room information, either room-cavity ratios or room areas.

Room-cavity ratios are based on rectangular or square rooms from 8 ft by 8 ft (2.4 by 2.4 m) to 100 ft by 300 ft (30.5 by 91.5 m) with cavity depth of 1 to 30 ft (.3 to 9.1 m). Hand or computer calculations must be done for different shaped rooms. Cavity depth is considered the distance from the light source to the work surface. Quick Calculation charts with room-cavity ratios show three curves, 1, 5, and 10. Do not be fooled by guessing room-cavity ratios. Room configurations profoundly affect ratio results. Wall surfaces absorb light. Consequently, with ceiling-mounted lights, the walls in high-ceiling narrow rooms absorb more light than in low-ceiling wide rooms. Therefore, the light absorbed by the walls does not get a chance to reach the work surface. In general, a room-cavity ratio of 10 usually means that the room's width equals the mounting height (the height between the surface lighted and the light source). A ratio of 5 usually means a room width 2 times the mounting height. And a ratio of 1 means the room width is 10 times the mounting height. These dimensions are not true in all cases; they are general guides.

Some charts do not identify room-cavity ratios, but identify room area. All the charts assume good room reflectances usually 80/50/20 (ceiling/walls/floor) unless otherwise noted. This information is suitable for quick comparisons of various luminaires. Often the information is not exactly the parameters of the design. To compensate, correction factors must be multiplied in and manufacturers supply the factors for different luminaires or different sources. (Additional parameters require hand- or computer-generated calculations.)

Use the charts by choosing either the amount of initial footcandles desired or spacing of the luminaires. Transfer the amount through the appropriate curve to the opposite axis to read the unknown quantity—footcandles if spacing is known, or spacing if footcandles are known.

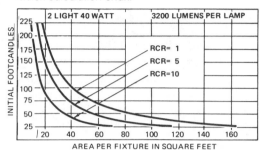

Quick calculation chart.

Lightolier

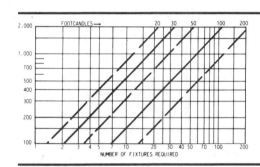

QUICK CALCULATOR CHART
1. Find room area on VERTICAL AXIS.
2. Move across chart to FOOT CANDLE CURVE.
3. Move down to HORIZONTAL AXIS and read NUMBER OF FIXTURES REQUIRED.

AE11X-T041
70/50/20 Reflectances
0.7 Maintenance Factor

Capri

Candlepower curve and tables: beam spread, coefficient of utilization, and spacing.

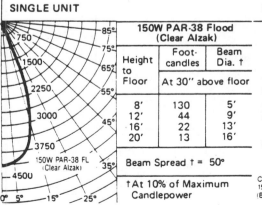

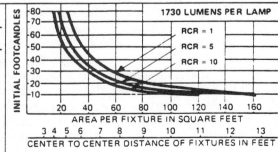

SINGLE UNIT

150W PAR-38 Flood (Clear Alzak)

Height to Floor	Foot-candles	Beam Dia. †
	At 30″ above floor	
8′	130	5′
12′	44	9′
16′	22	13′
20′	13	16′

Beam Spread † = 50°

†At 10% of Maximum Candlepower

MULTIPLE UNITS: 150W PAR-38 FL (CLEAR ALZAK) SPAC. RATIO = 0.7 E.T.L. NO. 392016

Room Cavity Ratio	% Effective Ceiling Cavity Reflectance								
	80			50			10		
	%Wall Reflectance								
	50	30	10	50	30	10	50	30	10
1	1.02	1.01	.99	.97	.96	.95	.90	.90	.89
2	.97	.94	.92	.92	.91	.89	.88	.86	.85
3	.93	.90	.87	.90	.87	.85	.86	.84	.83
4	.89	.85	.82	.86	.83	.81	.83	.81	.79
5	.85	.81	.78	.83	.80	.77	.80	.78	.76
6	.82	.78	.76	.80	.77	.75	.78	.76	.74
7	.80	.76	.73	.78	.75	.73	.76	.74	.72
8	.77	.73	.71	.76	.72	.70	.74	.71	.70
9	.74	.70	.68	.73	.70	.67	.71	.69	.67
10	.71	.67	.65	.70	.67	.64	.69	.66	.64

Conversion Factors:
150W and 250W PAR-38 (Gold and clear): C.U. x 1.0:
(Black multigroove): C.U. x 0.9

Conversion Factors:
150W PAR-38 (Gold): FC. x 1.0 Black multigroove) FC x 0.9
250W PAR-38 (clear and gold): FC. x 1.8 Black multigroove: FC. x 1.6

Lightolier

OTHER INFORMATION

Photometrics for fluorescent luminaires offer additional information for interpretation. They indicate maximum average brightness, Visual Comfort Probability (VCP), and zonal lumens. Luminance ratios should not exceed 5:1 for maximum to average footcandles in view, both crosswise and lengthwise to the luminaire, if not square. (Rectangular fluorescent troffers—2 by 4's and 1 by 2's—have different brightness crosswise and lengthwise; squares—2 by 2's—do not.)

A VCP of 80 or more is necessary for comfort. Keep light flux (direct light) out of the 60-90 degree zone. It can cause direct glare or reflected glare from computers. If the above conditions are met, glare should not be a problem. Design must include quality considerations.

Photometrics for luminaires also indicate efficiency. It is the ratio of the lumens leaving the luminaire to the amount of lumens produced by the light source. It is indicated by percentage; the higher, the more efficient the luminaire. Design should include energy-efficiency considerations.

Information is available to assist designers in determining what light source to use, where to put the fixtures, and what intensity and spread of light can be expected for effectiveness, efficiency, and economy. Use them.

Hand Calculations

POINT-BY-POINT METHOD

The point-by-point method is suitable for non-uniform lighting. It can be used to calculate the illumination for any surface from any angle, but the surface must be defined in size and must not be too big. Surfaces are usually walls, tabletops, or objects. The point-by-point method calculates the footcandles falling on a surface. It is based upon the requirements that only one source be calculated at a time and that the light from the source must come to the surface directly, not reflected from somewhere else. To perform the calculations, four pieces of information must be known: the angle of the light beam from nadir (0° angle), the candlepower at that angle, the distance from the light source to the surface, and the position (perpendicular, at an angle, or parallel) of that surface relative to the 0° light beam. The position of the surface affects the amount of light. Surfaces perpendicular to the light beam receive the most light.

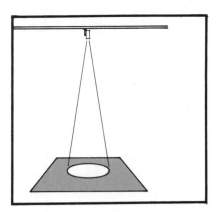

Point-by-Point Formula for a Surface Perpendicular to Nadir (0° angle)

$$\text{footcandles} = \frac{\text{candlepower}}{\text{distance}^2}$$

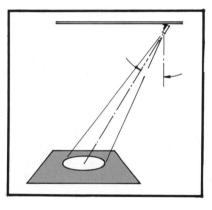

Point-by-Point Formula for a Surface at an Angle

$$\text{footcandles} = \frac{\text{candlepower at that angle} \times \text{cosine of angle}}{\text{distance}^2}$$

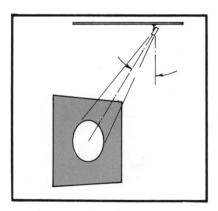

Point-by-Point Formula for a Surface Parallel to Nadir

$$\text{footcandles} = \frac{\text{candlepower at that angle} \times \text{sine of angle}}{\text{distance}^2}$$

Surfaces of light reprinted by permission of Lightolier.

An example of these hand-calculation methods is to determine what the illuminance from a 150-watt A-lamp downlight might be on the center of a picture on the wall (a surface parallel to nadir). The candlepower is determined from manufacturers' charts. (See *candle-power-distribution curves*.) At a 45° angle, it is 304. The cosine of the angle is determined from mathematic tables and is 0.7. The distance from the source to the picture is determined by drawing the triangle of the distances (distance down and distance away from) at scale and measuring the third leg of the triangle. In this case, the source is 3 ft (.9 m) out from the wall and the light hits the center of the picture 3 ft down. The third leg of the triangle measures 4 ft-3 in. (1.3 m). The calculations would be:

candlepower = 304 at 30° angle
sine of 30° angle = 0.7
distance = 4.24 ft (1.3 m)

$$\text{footcandles} = \frac{304 \times 0.7}{d^2} = \frac{212.8}{18} = 11.8$$

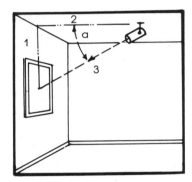

Triangle of distances
Aiming angle at (a)
(1) 3 ft   (2) 3 ft   (3) 4 ft-3 in.

If it is desirable to have the picture appear bright, it probably would not happen unless the surrounding surfaces received a lot less footcandles—not more than 4.

COSINE AND SINE TABLE FOR TYPICAL ANGLES

Angle	Cosine	Sine
30°	.866	.500
45°	.707	.707
60°	.500	.866

When more than one light source contributes light to the surface, each contribution at each angle and candlepower intensity must be calculated and added together. Hence, computers are magnificent for computing point-by-point measurements. Computers can quickly go through the necessary reiterations to arrive at the total.

ZONAL CAVITY METHOD

Zonal cavity calculations are suitable for uniform lighting and a whole space can be lighted with this method. This method can determine the number of light fixtures needed to provide a specified footcandle level uniformly. It can also be rearranged to determine how many footcandles will be received from a certain number of luminaires with a certain light-producing capability. As part of the calculation procedure, a room is divided into cavities. Minimally, there is always a room cavity. For instance, a room with recessed ceiling luminaires and a specific footcandle level on the floor has a cavity of the whole room. At the other extreme, a room with suspended luminaires and tabletop work surfaces with a specific footcandle level would have three cavities—ceiling cavity (from the lower edge of the luminaires to the ceiling), room cavity (from the lower edge of the luminaires to the tabletops), and floor cavity (from the tabletop to the floor). Calculations would be made for each cavity.

The zonal cavity method relies on ratios of room size (length, width), height of the cavity, room reflectances, and depreciation of the light over time (light loss). In this calculation method, light is expected to be reflected from the surfaces in the room as well as to be direct. Eight pieces of information must be known—the type of luminaire, the footcandle level, the room size, the room reflectances, the height of the cavity, a chosen light source, the amount of lumens from that source, and if possible, light loss factors.

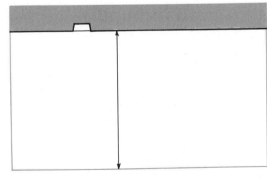

Room cavity zone only.

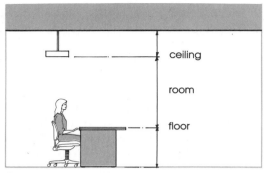

Cavity zones.

Light loss is caused by dirt in the room and the aging of the light source. Dirt in the room not only dulls reflectance possibilities of the major interior surfaces but also accumulates on luminaires. The design of luminaires makes some more susceptible to dirt deposits than others. The IESNA, *Lighting Handbook, Reference and Application, 1993* contains information on light loss factor. Further, light sources have different abilities to continue producing their lumens. Manufacturers publish such information as lamp lumen depreciation amounts.

With the zonal cavity method, determine the amount of footcandles needed by the amount of light required to do the task. Determine the room sizes and room reflectances. Determine the coefficient of utilization. Tables of coefficients are published by manufacturers. To use these tables, room size must be translated into a cavity ratio. The ratio is used to choose an appropriate coefficient with the appropriate room reflectance.

$$\text{cavity ratio} = \frac{5 \times \text{height of cavity} \times (\text{length of room} + \text{width of room})}{\text{length} \times \text{width}}$$

Determine the lumens per light bulb and how many bulbs are to be used in each luminaire. Lumens are published by manufacturers and are available in technical catalogs.

Zonal Cavity Formula

$$\frac{\text{number}}{\text{of fixtures}} = \frac{\text{footcandles desired} \times \text{total room area}}{\text{light bulbs per luminaire} \times \text{lumens per light bulb} \times \text{coefficient of utilization} \times \text{light loss}}$$

The zonal cavity method permits designers to light a space uniformly with specific footcandles on any horizontal surface, such as on a library tabletop or on the floor. Lighting is a sci-

ence as well as an art. Often, the science of lighting is underused. It is not difficult to master at least one calculation method. The effects of predetermining and designing lighting at specific levels are powerful. Do not ignore them. Lighting actually can be more artful when scientifically designed.

Computer-Generated Calculations

Computers can manipulate a lot of information for lighting design. Personal computers must have powerful hardware, including a math coprocessor, suitable software and preferably a color monitor. The software is produced by electrical utility companies (for example, Southern Cal Edison), by fixture and light-source manufacturers (Globe, Halophane, GE, etc.), by universities (MIT), by research organizations (Lawrence Berkeley Lab), and by private firms (Lighting Technologies, RLH Lighting Consultants, Lighting Sciences, etc.). Not all of the software is the same. Some calculate daylight and some electric light, indoors with interreflections or outdoors without. Some crunch numbers for lighting analysis and some paint pictures for visualizing lighted spaces. Some analyze life-cycle cost of a system. Manufacturers' software is limited to calculations using their products.

Input can be digital or graphic. Daylight calculations require input of space size, shape, reflectances of surfaces, and orientation to the sun. Electric light calculation input depends upon the output desired. Input includes space (size, shape, reflectances), luminaires (photometrics, mounting height, layout, aiming angle, aiming point), use (power costs, operating hours), and so forth.

Output can be experimentation to develop a plan or evaluation to test a plan. Experimentation output can be used in the initial phases of design for visualization. Image output helps the visualization process for both the designer and the client. It is a talking picture and leads to insights about the design. The designers can visualize and communicate the design. Likewise, the output can be used for client presentations. Evaluation output can be used in the determination phase (lighting performance and luminaire selections), for assuring suitable amounts of light, and used in the production phase (electrical plans and specifications). Numerical output is always X, Y coordinates. Sometimes it is selected points or areas.

Lighting performance and luminaire selection information can be photometric or lamp data based. Calculations performed can be point-by-point, zonal cavity, and life-cycle/energy determinations. Unlike hand-generated calculations, some computer-generated point-by-point calculations take into consideration interreflectances from adjacent surfaces, an impossible task for hand calculations. Number crunching computers eat it up. For outdoor applications, computations for size of overhangs, fins, and louvers for daylight reduction can be generated. Likewise, some software calculates performance of both daylight and electric light together—a boon for a difficult engineering task.

The electrical plans and specifications for the production end of lighting design once entered into the computer can easily be changed as the job changes (when the heat/air contractor changes duct work and eliminates some recessed fixtures spaces). Computer changes are easy to spit out and are hard to ignore.

All in all, the time invested learning and using the computer pays off, but the information put out is only as good as the information put in. The computer adds only speed of computation. You are still the judge and your client is still the jury.

The output can be calculations or graphics. The calculation output can be:

- Illuminance—task, horizontal and vertical surfaces both flat and convex (front and sides, developed for sports lighting), other room surfaces, and visual comfort probability.
- Luminance—task background, horizontal and vertical surfaces, Relative Visual Performance, and luminance ratio.
- Power budgets.

The graphic output can be:

- Ray traces (almost like photographs)
- Radiosity
- Simulated lighting distributions
- Solid modeling

Overall, graphics reduce the number of design alternatives that need further studying. They could substitute for building lighted models for some jobs. Graphics let the designer gain insights into the visual quality of lighting design proposals. Calculations let the designer gain insights into the quantity and measurable quality of lighting proposals. Unfortunately, not all qualities (sparkle, shadow, highlight, and color enhancement) are included in computer outputs. These qualities need special attention.

A library of photometric data is required for effective computer design. If the designer is limited to minimal choices of photometric data, then the designs will be limited. In addition, the designer must be familiar with the logic of lighting computations in order to interpret the output correctly and to identify errors. Computers save lengthy computations, solving as many as 4,000 simultaneous equations. Likewise, once the information is in the computer, it is easy to change. Changes are probably the greatest payback for the time and effort the professionals—architects, interior designers, and contractors—put into computerization. Changes for the computer are a snap.

Video cameras calibrated to see light like the human eye can put data into computers and can print out anticipated visual performance, based on visibility, brightness contrast of the task and overall brightness level. This computer system goes beyond the standard number crunching from photometric data.

lighting design

The goals of architectural lighting are to illuminate the structure, its contents, the tasks that go on there and to reflect on people. (On the other hand, the goal of theater lighting is to illuminate people directly.) Architectural lighting should enhance the visual appearance of the structure and its contents. But also, it should be visually comfortable and give the luminance needed for people to do their tasks. Measure it in terms of visibility, not raw footcandles. Visibility relies on the viewer's visual acuity, the reflectance of surfaces, and the amount, direction, and adjacent brightness of light. Finally, architectural lighting should reflect flatteringly on people. For example, light reflected from vertical surfaces can be flattering. Light coming directly from over-the-head can be unflattering.

With these goals in mind, remember that

light can only be seen on a surface and/or at a visible source. The end product of lighting design is visual. Consequently, design lighting visually. Determine how the space should look:

- What surfaces should be bright?
- What, if any, sources should be seen?

Visual design can be accomplished by sketches (if you are able), by *Rub & Show Lighting Graphics* from this publisher (if you do not have time or are not able to sketch), or by computer simulations. Do not miss this important first step.

Second, design lighting mathematically. The mathematics involved in lighting are basically addition, subtraction, multiplication, and division—all taught in high school. For example:

- Addition of interreflections of multiple sources,
- Subtraction of surface color absorption,
- Multiplication of criteria, of conversion factors, or to obtain squares,
- Division of ratios.

The rest, square roots, cosines, etc., can be looked up in the mathematic tables before adding, subtracting, multiplying or dividing. (However, the ability to visualize three-dimensionally makes the mathematics more understandable. That skill is not available in tables, but is learnable.)

To design mathematically, determine how much light is needed on the surfaces. For instance, determine the amount of light needed to reveal surface colors pleasantly. Remember, surface color can be enhanced only if the color is contained in the light. Remember, the eye is more sensitive to yellow-green rather than red or blue. Hence sources with a large yellow-green part of the spectrum preserve surface colors better when dimmed. In addition, sources that have a broad spectrum of color make multiple surface colors look better. Remember that up to a point, the greater the amount of light, the more the colors are revealed. But, too much light makes colors appear overexposed (like overexposed photographs). On the other hand, too little light makes colors appear dim (like underexposed photographs). In general, fluorescent sources have the greatest range of footcandles to make colors appear natural, from about 12 to 1,000. Incandescent has a narrow range, from about 8 to 50.

Remember, when using Kelvin degrees (K) for choosing sources, the higher the K, the bluer the light appears; the lower the K, the redder. Likewise, when using color rendering index (CRI), remember the higher the index, the less color distortion; the lower the index, the greater distortion.

Remember that 3,500 degrees Kelvin (35K) fluorescent reveals colors well, but does not enhance black-and-white reading tasks. Black-and-white contrast is enhanced by T-8 in 30K at 95 CRI and T-12 in 50K at 92 CRI.

To complement the color of space with the color of the light, choose according to the amount of light designed for the space. If the amount of light is to be low, use warm sources (28 to 35K) either fluorescent, incandescent, or white sodium. If amount is designed to be high, use cool sources (35 to 51K), either fluorescent, halogen low-voltage, or metal-halide.

Further, use amount, Kelvin degrees, and Color Rendering Index together to determine how well colors will appear in a space. Thus, if interior color scheme will be cool, warm it up with warm sources (not 41K), particularly if the amount of light will be low. If interior color scheme will be warm, cool it down with cooler sources, like halogen, low-voltage, and 35K fluorescent.

Remember, when using remote transformers for low-voltage lighting, electricity drops as it travels along the wire. The drop affects color and amount of light. Keep voltage drop to 5 percent.

Determine what amount of light is best for a task by the age of the person, task size and contrast, and speed and/or accuracy required. (Consult the IESNA *Lighting Handbook, Reference & Application*, 1993.) Remember that, among other things, the best possible light for a visual task depends partly upon where the light is coming from. Light from an angle that creates reflected glare reduces task contrast. Task contrast should be as great as possible. But contrast (luminous) ratios within a space should not be great. Keep within the recommendations for the best possible visual environment.

LUMINANCE RATIOS FOR NON-COMPUTER TASKS

AREA	RANGE OF LIGHT RELATED TO TASK LUMINANCE
Close to task	Not greater than task Nor less than $\frac{1}{3}$ of task
Remote from task	Not greater than 10 × task Nor less than $\frac{1}{10}$ of task

Adapted from IES Lighting Handbook, 1987 Application Volume.

LUMINANCE RATIO FOR COMPUTER TASKS

AREA	LUMINANCE
Close to task	Brighter surroundings not greater than 3 × task
	Darker surroundings not less than 1/3 of task

Adapted from IES Lighting Handbook, 1987 Application Volume.

CONTRAST RATIO FOR COMPUTER TASKS

AREA	CONTRAST
Monitor image to background	at least 10:1

Reprinted by permission of Illuminating Engineering Society (See Sonnefield in Bibliography).

Likewise, determine that the task light can be installed so that it will not create source glare (direct) or surface glare (reflected). (See Chapter 19.)

For instance, determine what luminance ratios are needed to visually emphasize architecture, interior finishes, or merchandise. A ratio of 3:1 can call attention, but 17:1 will command attention. Within any space where long-term tasks go on, keep luminance ratios of surfaces within view at a ratio to 5:1 for visual comfort (but 3:1 is better). The luminance ratio for examining merchandise should not be more than 3:1 between where merchandise is displayed (racks, etc.) and where merchandise is appraised (dressing room). Finally, the uniformity or luminance ratio of a ceiling surface indirectly lighted in a large space should be 10:1 or less.

When determining the amount of light, remember that the light coming from an angle will be less than if it came perpendicular to the surface. Also, remember that surface color, even white, will reduce light by absorption. Multiply surface reflectance (LR%) times the amount striking the surface to find what the level will be. No color or material has 100 percent reflectance. (Manufacturers wish that they did.) On the other hand, mirrors reflect the highest (80 to 99 percent). Glass surfaces reflect at the same time as they transmit. How much they reflect depends upon the light levels on both sides of the glass. It always reflects some. At night, windows reflect like translucent mirrors. Hence, never light mirrors or glass with direct light. For example, never use open downlights over a glass-topped table. The direct light creates reflected glare. It is obnoxious and inescapable. Avoid glare at all costs.

At the same time, remember that only direct light creates shadow and sparkle. Shadows permit us to see depth and form. Texture is made visible with direct grazing light creating shadows. (See Chapter 5.) Shadows are necessary for a well-defined visual scene. Equally important, sparkle is necessary for a vibrant visual scene. Surfaces must be glossy to reflect sparkle. Large glossy surfaces produce glare. Sparkle is small. Without sparkle and shadows, spaces appear dull and gloomy, like a drab, overcast day.

Third, determine if you want visible sources to show light. Remember that visible fixtures (chandeliers, sconces and pendants) are lighting jewelry. They should not be expected to significantly contribute light for a space. To the contrary, using chandeliers requires that the ambient light be high enough to wash out the brightness effect of the bare sources. Even so, use lighting jewelry whenever it is affordable. Jewelry is appreciated by almost everyone. Specify low wattage for bare bulbs and reduce the direct-glare impact. Likewise, remember that surface-mounted or recessed fixtures with lenses and diffusers will be visible sources. For visual comfort, specify lenses and diffusers that will be low in brightness. Nonetheless, research at Penn State has pointed out that people felt that the space tested appeared 20 percent brighter when a small part of the lighted source (a narrow low-brightness lens on a fluorescent fixture) was visible as compared to the space with sources hidden. The low-bright light directly from a source makes an impact.

Fourth, choose sources, fixtures and con-

trols. Determine what light source can provide the light needed (amount and distribution). Use interpretive and hand- or computer-generated calculations. Ceiling height profoundly affects light-source choice. For instance, low-voltage sources require a 10 ft (3 m) or less ceiling height to be effective. Metal-halide requires 10 ft or more. Light source needs to be the first choice. Fixtures are the second.

Determine what fixture can enclose the light source, provide the distribution needed, and fits the architectural (and budgetary) constraints. Use interpretive calculations, specifications, and price sheets provided by manufacturers. Verify and expand your interpretive calculations with hand- or computer-generated calculations.

Determine whether the choices of sources and fixtures are the most energy efficient. In all situations (task, accent, wall-wash, and ambient), use the most energy-efficient source and fixture. In general, use compact fluorescent instead of incandescent where point-source, high-intensity light is not needed. Use metal-halide, low-wattage white sodium, or single-source rim lighting instead of a ceiling full of fluorescent fixtures. Likewise, for architectural lighting,

use prismatic tube or fiber optic, single-source lighting. Use low-wattage, low-voltage incandescent instead of high-wattage incandescent for point-source applications. Check both source and fixture efficiency listed in manufacturers' technical data. Save our unrenewable resources and provide good light at the same time. Light-source manufacturers are eager to assist in choosing efficient sources for large installations. For businesses, energy-efficient sources operate 5 days per week for the whole year. The energy savings pays back the extra cost of the more expensive light source and fixture, usually within one to two years. After that, the energy savings becomes a cash savings for the life of the building or the life of the company, whichever comes first. Do not let your clients be fooled by front-end higher costs. Sell the long-run lower costs.

Sample Electric Cost
Substitute an 18-watt compact fluorescent source for a 75-watt A incandescent source in countertop table lamps in retail stores and save 57 watts for 4,000 hours per year and cut $22.80 per table lamp off of the utility bill.

Reprinted by permission of Tribune Media Services.

Next, determine the controls for the lighting system. Controls permit reduction of energy consumption when space is unoccupied or when daylight is available. In addition, controls provide for different lighting effects.

SAVINGS DUE TO DIMMING FOR
INCANDESCENT LIGHTING

% DIMMED	ELECTRICITY SAVED	EXTENDS SOURCE LIFE
10	5%	2 times
25	10%	4 times
50	25%	20 times
75	50%	+20 times

Reprinted by permission of Lutron.

The appropriate size for a control can be determined by totalling the number of watts to be dimmed and by choosing the dimmer with a 20 percent larger capacity. If using low-voltage sources, specify either a magnetic or an electronic transformer with the suitable dimmer. If using fluorescent, specify a dimmable ballast.

Designers can determine position, wattage, and aiming angle of light sources by manipulating the known variables and finding the unknown. Interpretive calculations from manufacturers are suitable for obtaining a great deal of information. For example, when using Lamp- or Lighting-Performance Data and lighting a piece of art covered with glass, the known variables are: the ceiling height, how far down the wall the artwork will be hung, the size of the art, the intended footcandles and the aiming angle. (A steep angle is needed to avoid reflected glare. Therefore, the angle must be 60 degrees, with 0 as perpendicular to the wall.) Then, determining the unknowns: the position on the ceiling and the light source is not difficult. They are obtainable from the published data.

On the other hand, if the known variables are: position of the fixture, light source and the aiming angle, by using the same Lamp- or Lighting-Performance Data, the unknown variables: beam location, size, and footcandles can be determined.

Refine mathematics of lighting design with hand- or computer-generated calculations. Refinement for both essential and luxurious lighting has no substitute.

Fifth, check luminance ratios to assure that brightness will be pleasing, flattering, and not fatiguing to those who occupy the space. Controls can help to refine the lighting design and obtain the desired brightness and lighting balance. Balance is a mark of a good lighting design, unless the objective is to create tension and excitement. Balance pleases the occupants. Balance is under applied and overly needed.

Sixth, meet building codes; always local, sometimes state, and always national. Some states limit the amount of energy consumption allowable for specific spaces. Further, local and state fire safety boards regulate lighting. Be certain that specifications and installation comply with all applicable regulations. They are developed for protecting unrenewable resources and for safety.

Even though we adapt to most amounts of light (a little or a lot) without eye pain and even though we do not miss light unless it is completely gone, light affects our ability to complete activities effectively, to see colors and textures, and to appear "in-the-best light." Likewise, light adds visual value and calls attention to surfaces. It has long-lasting importance. Once installed, light fixtures remain for the life of the building or until the lighting is remodeled. Dirt on fixtures reduces the amount of light. Therefore, maintenance is important. Use does not damage lighting fixtures. Abuse does. Unlike kitchen countertops or bathroom faucets, which do wear out with use, lighting fixtures wait to be replaced or to be removed by the wrecking crew. Consequently, thoughtfully specify lighting. It will be there a very long time.

Reprinted by permission of King Features Syndicate, Inc.

bibliography

Anderson, Beth. "Boardroom Shakeup." *Audiovisual Communications*, 22, no. 11 (November 1988).

Armstrong, Tim. *Colour Perception*. Norfolk, England: Tarquin Publications, 1996.

Arnheim, Rudolf. *Art and Visual Perception*. Berkeley: University of California Press, 1974.

Bell, Simon. *Elements of Visual Design in the Landscape*. London: E & FN Spon, 1993.

Benya, James R. "The Lighting Design Professional." *Architectural Lighting*, 2, no. 1 (January 1988), no. 2 (February 1988), no. 3 (March 1988), no. 4 (April 1988).

Bernecker, Craig A. "Designing for People: Applying Psychology of Light Research." *Architectural Lighting*, 1, no. 10 (November 1987).

Boylan, Bernard R. *The Lighting Primer*. Ames: Iowa State University Press, 1987.

Brown, G.Z. *Sun, Wind and Light: Architectural Design Strategies*. New York: John Wiley & Sons, 1985.

Burnie, David. *Light*. New York: Dorling Kindersley, 1992.

Butler, H., ed. *Home Decorating Using Light*. London: Marshall Cavendish, 1974.

Butterfield, Jan. *The Art of Light + Space*. New York: Abberville Press, 1993.

Corth, R. "Human Visual Perception." *Lighting Design and Application*, 17, no. 7 (July 1987).

DiLaura, D.P., Igoe, D.P., Samaras, P.G., and Smith, A.M. "Verifying the Applicability of Computer Generated Pictures to Lighting Design." *Journal of the Illuminating Engineering Society*, 17, no. 1 (Winter 1988).

Dondis, Donis A. *A Primer of Visual Literacy*. Cambridge: Massachusetts Institute of Technology, 1973.

Dulanski, Gary. "Dimming: Another Side of Light Control." *Lighting Design and Application*, 17, no. 12 (December 1987).

Dulanski, Gary. "Managing Energy with Lighting Controls." *Lighting Design and Application*, 17, no. 11 (November 1987).

Durrant, D.W., ed. *Interior Lighting Design*. London: Lighting Industry and the Electricity Council, 1973.

Effron, Edward. *Planning and Designing Lighting*. Boston: Little, Brown & Co., 1988.

Egan, M.D. *Concepts in Architectural Lighting*. New York: McGraw-Hill, 1983.

Electricity Council. *Better Office Light*. EC

2873R. London, 1972.

Eley Associates. *Advanced Lighting Guidelines*: 1993. Palo Alto: Electric Power Research Inst., 1993.

Elmer, W.B., and Lemons, T.M. "Indirect Rim Lighting for Building Interiors." *Lighting Design and Application*, 17, no. 7 (July 1987).

Evans, Nancy. "Lighting for Safety and Security." *Light Magazine*, 39, no. 2 (1970).

Fahsbender, Myrtle. *Residential Lighting*. New York: Van Nostrand, 1947.

Florence, Noel. "The Energy Effectiveness of Task-Oriented Office Lighting Systems." *Lighting Design and Application*, 9, no. 1 (January 1979).

Flynn, John F., "A Study of Subjective Responses to Low Energy and Nonuniform Lighting Systems." *Lighting Design and Application*, 7, no. 2 (February 1977): 6-15.

Flynn, John F., and Mills, Samuel M. *Architectural Lighting Graphics*. New York: Reinhold, 1962.

Freeth, Richard. *Plan Your Home Lighting*. London: Studio Vista, 1970.

Gardner, C., and Hannaford, B. *Lighting Design, An Introductory Guide for Professionals*. New York: John Wiley & Sons, 1993.

Gardner, Robert. *Investigate and Discover Light*. Englewood Cliffs: Simon Schuster, 1991.

General Electric Company. *Store Lighting*. Lighting Application Bulletin. Cleveland, 1993.

General Electric Company. *Industrial Lighting*. Technical Paper 108R. Cleveland, 1977.

General Electric Company. *The Light Book*. Technical Paper. Cleveland, 1982.

General Electric Company. *Office Lighting*. Lighting Application Bulletin. Cleveland, 1992.

General Electric Company. *Optical Fiber Illumination Systems*. Lighting Application Bulletin. Cleveland, 1995.

General Electric Company. *Specifying Light and Color*. Lighting Application Bulletin 205-61701. Cleveland, 1986.

General Electric Company. *Stores*. Lighting Application Bulletin 205-51701. Cleveland, 1987.

Gilliatt, Mary, and Baker, Douglas. *Lighting Your Home: A Practical Guide*. New York: Pantheon, 1979.

Gordon, Gary and Nucholls, James. *Interior Lighting for Designers*. New York: John Wiley & Sons, 1995.

Green, William. *The Retail Store*. New York: Van Nostrand Reinhold, 1991.

Grief, Martin. *The Lighting Book, A Buyer's Guide to Locating Almost Every Kind of Lighting Device*. Pittstown, NJ: Main Street Press, 1986.

GTE Electrical Products. "Lighting Visual Display Terminal Areas." *Engineering Bulletin* 0-364. Danvers, MA: Sylvania, 1986.

Gudum, J. "Semicylindrical Illuminance in Sports Lighting." *International Lighting Review*, 2, 1984.

Halo Lighting Division. *The Language of Lighting*. Elk Grove Village: McGraw-Edison, 1983.

Harman, T., and Allen, C. *Guide to the National Electric Code*. Englewood Cliffs: Prentice Hall, 1990.

Hebert, P. "Matching Color and Light in Interior Design." *Lighting Design and Application*, 17, no. 9 (September 1987).

Helms, Ronald N. *Illumination Engineering for Energy Efficient Luminous Environments*. Englewood Cliffs: Prentice-Hall, 1991.

Hockey, S. Newton. "Piping Light." *International Lighting Review*, 2 (October 1985).

Hopkinson, Ralph G., and Kay, John D. *The Lighting of Buildings*. London: Faber and Faber, 1972.

Illuminating Engineering Society of North America, Daylight Committee. "Recommended

Practice of Daylighting." *Lighting Design and Application*, 9, no. 2 (February 1979).

Illuminating Engineering Society of North America. IESNA *Lighting Ready Reference*. New York, 1996.

Illuminating Engineering Society of North America. *Lighting Handbook, Reference and Application*. New York, 1993.

Illuminating Engineering Society of North America, Merchandising Lighting Committee. *Recommended Practice for Lighting Merchandising Areas*. New York: Illuminating Engineering Society of North America, 1986.

Jankowski, Wanda. *Designing with Light, Residential Interiors*. New York: PBC International, 1991.

Jankowski, Wanda. *The Best of Lighting Design*. New York: PBC International, 1987.

Jones, Gerre. *How to Market Professional Design Services*. New York: McGraw-Hill, 1973.

Kalff, Louis C. *Creative Light*. New York: Van Nostrand, 1970.

Kambich, Dave G. and Rea, Mark S. "New Canadian Lighting Analysis System." *Lighting Magazine*, 1, no. 9 (November 1987).

Kamm, Dorothy. "Color Renditions: Light as Decoration." *Lighting Design and Application*, 15, no. 12 (December 1985).

Kamm, Lloyd J. *Lighting to Stimulate People*. Boston: Christopher Publishing House, 1948.

Kimsey, S.P., ed. "Comparative Analysis of Alternative Energy Efficient Lighting Systems." *Architecture*, 77, no. 6 (June 1988).

Kodak, Motion Picture and Audiovisual Division. *Designing for Projection*. V3-141. Rochester: Kodak, 1977.

Lam, William. *Sunlighting as Formgiver for Architecture*. New York: Van Nostrand Reinhold, 1986.

Lam, William. *Perception and Lighting as Formgivers to Architecture*. New York: McGraw-Hill, 1977.

Lawson, Bryan. *How Designers Think; The Design Process Demystified*. London. Buttersworth Architecture, 1990.

Liberman, Jacob. *Light: Medicine of the Future*. Sante Fe: Bera & Company, 1991.

Lightolier. *Designing with Light*. Application Guide. Jersey City: Lightolier, 1996.

Lightolier. *Notes on the Lighting of Walls, Pictures, Draperies, and Other Vertical Surfaces*. Application Guide 10.02. Jersey City, 1974.

Linton, H. *Color Model Environments: Color and Light in Three-Dimensional Design*. New York: Van Nostrand Reinhold, 1985.

Lord, David. "Simulation of Lighting Design." *Architecture*, 77, no. 6 (June 1988).

Maybin, Harry B. *Low Voltage Wiring Handbook*. New York: McGraw-Hill, 1995.

McCloud, Kevin. *Lighting Book*. London: Ebury Press, 1995.

Moyer, Janet L. *The Landscape Lighting Book*. New York: John Wiley & Sons, 1992.

National Lighting Bureau. *Lighting Energy Management in Retailing*. Washington, DC, 1982.

Orfield, S.J. "Open-Plan Office Lighting." *Lighting Design and Application*, 17, no. 7 (July 1987).

Ortho Books. *"How to Design and Install Outdoor Lighting."* San Ramon, CA: Chevron, 1984.

Pankin, Sidney M. "The Parts Department." *Architectural Lighting*, 1, no. 6 (June 1987).

Pelger, Martin. *The Dictionary of Interior Design*. New York: Bonanza, 1966.

Philips Lighting Company. *Education Facility*, Application Guide. Somerset, NJ, 1995.

Philips Lighting Company. *Office Lighting,*

Application Guide. Somerset, NJ, 1992.

Philips Lighting Company. *Retail Lighting*, Application Guide. Somerset, NJ, 1991.

Philips Lighting Company. *Security Lighting*, Application Guide. Somerset, NJ, 1992.

Phillips, Derek. *Planning Your Lighting*. London: Design Council, 1976.

Robbins, Claude L. *Daylighting: Design and Analysis*. New York: Van Nostrand Reinhold, 1986.

Rosenberger, Paul. *The Complete Electronic House*. Englewood Cliffs: Prentice Hall, 1991.

Rossbach, Sarah. *Interior Design with Fegn Shui*. New York: Penguin Books, 1987.

Smith, Fran K., and Bertolone, Fred J. *Bringing Interiors to Light*. New York: Whitney Library of Design, Watson-Guptill, 1986.

Sonnenfield, S., Bonsignore, S., Howard, G.T., Nissen, R.J., and Clark, C. "Lighting for Teleconference Rooms." *Lighting Design and Application*, 15, no. 5 (May 1985).

Sorcar, Prafulla C. *Architectural Lighting for Commercial Spaces*. New York: John Wiley, 1987.

Sorcar, Prafulla C. *Energy Saving Lighting Systems*. New York: Van Nostrand Reinhold, 1982.

Sorcar, Prafulla C. *Rapid Lighting Design and Cost Estimation*. New York: McGraw-Hill, 1979.

Steffy, Gary R. *Architectural Lighting Design*. New York: Van Nostrand Reinhold, 1990.

Steffy, Gary R. *Lighting the Electronic Office*. New York: Van Nostrand Reinhold, 1995.

Sudjic, Deyan. *The Lighting Book: A Complete Guide to Lighting Your Home*. New York: Crown Publishers, 1985.

Sunset Books. *Home Lighting*. Menlo Park: Lane Publishing, 1982.

Szenasy, Susan S. *Light: The Complete Handbook of Lighting Design*. Philadelphia: Running Press Book Publisher, 1986.

Taylor, Joshua C. *Learning to Look*. Chicago: University of Chicago Press, 1981.

Turner, Janet. *Lighting*. London: B.T. Batsford, 1994.

Van Gills, A.M.F. "The Four-Corner Philosophy: A Matrix for Shop Lighting." *International Lighting Review*. 2nd Quarter, 1989.

Wallaschlaeger, Charles, and Busic-Snyder, Cynthia. *Basic Visual Concepts and Principles*. Dubuque, Iowa: Wm. C. Brown Publishers, 1992.

Watson, Lee. *Lighting Design Handbook*. New York: McGraw Hill, 1990.

Wells, Stanley. *Period Lighting*. London: Pelham, 1975.

White, Edward T., and Rubin, Arthur I. *Tracing Lighting Design Decisions for Open Office Space: A Pilot Study*. Washington, DC: National Bureau of Standards, 1981.

Whitehead, Randall. *Residential Lighting, Creating Dynamic Living Spaces*. Washington, DC: AIA Press, 1993.

Williams, H.G. "Office Lighting and Energy." *Skyscraper Management*. Washington, DC: Building Owners and Managers Association Institute (April 1975).

Willis, Lucy. *Light: How to See It; How to Paint It*. Cincinnati: North Light Books, 1991.

Wilson, A. "Achieving Energy Efficient Lighting." *Architecture*, 77, no. 6 (June 1988).

Yee, Roger. "The New Lighting: Too Dim and Too Dull?" *Corporate Design & Realty*, 5, no. 6 (June 1986).

Zackrison, Harry B., Jr. "Outside Lighting Systems Design." *Lighting Design and Application*, 10, no. 5 (May 1980).

index